THE HOMEOWNER'S PEST CONTROL HANDBOOK

2nd Edition

Warning and Disclaimer

THE HOMEOWNER'S

PEST

CONTROL HANDBOOK
2nd Edition

Gene B. Williams

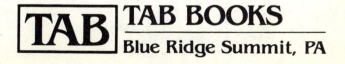

TAB **TAB BOOKS**
Blue Ridge Summit, PA

To My Wife
CYNTHIA
And our son
DANIEL

SECOND EDITION
SECOND PRINTING

©Copyright 1989 by **TAB BOOKS**.
TAB BOOKS is a division of McGraw-Hill, Inc.

Printed in the United States of America. All rights reserved. The publisher takes no responsibility for the use of any of the materials or methods described in this book nor for the products thereof.

Library of Congress Cataloging-in-Publication Data

Williams, Gene B.
The homeowner's pest control handbook / by Gene B. Williams.—2nd ed.
p.   cm.
Rev. ed. of: The homeowner's pest extermination handbook. c1978.
Includes index.
ISBN 0-8306-1139-8   ISBN 0-8306-3139-9 (pbk.)
1. Household pests—Control.   I. Williams, Gene B. Homeowner's pest extermination handbook. II. Title.
TX325.W53   1989
648'.7—dc20   89-31833
                                                                              CIP

TAB BOOKS offers software for sale. For information and a catalog, please contact TAB Software Department, Blue Ridge Summit, PA 17294-0850

Questions regarding the content of this book should be addressed to:

Reader Inquiry Branch
TAB BOOKS
Blue Ridge Summit, PA 17294-0850

Acquisitions Editor: Kimberly Tabor
Technical Editor: Lori Flaherty
Production: Katherine G. Brown

Contents

Introduction

There are over one million different species of animal on our planet. More than 70 percent (about 715,000) are insects, and about 10,000 new species of insect are discovered every year. This doesn't include other pests, like spiders, scorpions, and centipedes, none of which are insects.

Many species have only a few thousand members; others have populations in the billions. Individual ant or termite colonies often contain more members than do some large cities.

The study of insects is called entomology. Entomologists are, perhaps, one of the least appreciated of all scientists. After all, who would be interested in bugs? Who other than a child could find them so fascinating that he would spend hours watching them and dissecting them?

Bugs aren't all that interesting, are they?

At first, this may seem the case to many people. But the insect world is one of the most interesting and varied of the entire animal kingdom. Nowhere else will you find such diversity. Insects do things you would never suspect them capable of doing.

They are also of extreme economic importance. Crop damage from insect pests runs into billions of dollars every year. Termites in the United States cause nearly ten times the amount of damage of fires and storms combined.

Other insects, such as bees, pollinate crops. Without them, humanity would be in pretty bad shape.

Insects are found wherever man is, and a few places where man can't survive. From the highest mountains to the deepest valleys, from wet tropical jungles to burning deserts or the icy cold of snowbound areas, there are insects. In size, they vary from the microscopic to huge monstrosities reaching lengths of almost two feet. Colors vary like the rainbow, from earthy browns to bright, metallic greens, blues, and reds.

They crawl, jump, fly, and swim. Some are carried around by willing (and unwilling) hosts. They make their homes in plants, trees, the air, the ground, or inside animals. Their food consists of just about anything you can imagine.

When we think of pests in the home we immediately think of insects, spiders, and mice. There are thousands of pest control companies in America, making several billion dollars every year killing the tiny crawlers.

Obviously, pest control is a big business. In fact, it is one of the largest in the world. So large that there is even a National Pest Control Month.

So, one morning you wake up to find some horrible creature creeping across the table. Out comes the ever-present aerosol can of insecticide. But, try as you might, the problem not only continues, but gets worse.

Not only that, but you notice that something is eating everything in your garden.

What to do, what to do?

You flick through the Yellow Pages, and pick one of the hundred-odd companies listed there. What kind of price do they demand to come out and kill these darned bugs?

"Well, we can do it for a fee of $65, after which we'll come by to continue control for only $25 a month."

Figure it out. That's $340 for the first year, plus whatever tax is applicable.

And, $300 a year after that.

With a pained expression you reach for the checkbook to see if the budget will stand the strain this month, or even this year. As you flip open the checkbook, a silverfish whose supper you've just interrupted gives you a dirty look and scampers off.

"Can I get it just one time?" (That $65 price still hurts.)

"Sure. But, there'll be no guarantee with it."

Desperation works its way into your voice as you inform the salesman that you'll let him know. Then, you drive to the local library to find a book on how to control the bugs yourself.

None. (Well, one. This one.)

Of course, you could try the "home remedies" your friends and neighbors suggest. But, do they really know any more about it than you do? If the remedy doesn't work, what do you do then?

Well, you could fork over $340, and cry a lot. Or, . . .

You can use this book.

Everything you need is right here. The equipment you'll need, how to use it, and where to buy it, are all covered in Chapter 1.

Want to know something about any of the chemicals you'll be using? Refer to Chapter 2. There isn't any great mystery to them. If you follow the simple directions, they can be perfectly safe. Chapter 2 tells you what they are, what they'll do, how and where to use them, and for which pest. It's all listed alphabetically for easy reference.

Chapter 3 tells you of the best-known method of pest control, that of home maintenance. The easiest way to control pests is to not let them get established in the first place. Chapter 3 will tell you exactly what to do, and why.

Chapter 4 takes you through the basic exterminating procedures.

Chapter 5 gives you some of the basic information on the various pests you might encounter.

Chapter 6 goes into detail on each individual pest. Most of the pests you'll run into are listed in alphabetical order. Nothing could be easier. Along with specific information on each pest, you'll be told *exactly* what to do about it.

The final chapter wraps everything up with information on the Environmental Protection Agency, Poison Control Centers, and a few final remarks to make your task easier and safer. At the back of the book there is even a place where you can record important phone numbers in case of an emergency.

1

Equipment

Your first consideration is what equipment to buy. This will depend largely on your particular circumstances, and the forms of chemical (liquid, dust, etc.) you intend to use. This last consideration is usually determined by the pests you have and where the chemical will be applied.

Following the usual rule of thumb, get the best equipment you can afford. A $5 spray tank will do the job, but it won't do it efficiently, and probably not accurately. It will also tend to break down more often, and give out completely, sooner than the more expensive equipment.

If you plan on doing your own pest control from now on—and you should—the investment in better equipment will pay for itself in the long run. Good equipment, properly cared for, can last a lifetime. Or two.

Most places that carry chemicals will also carry various types of equipment. Nurseries, garden shops, hardware stores, do-it-yourself stores, and even some drug and department stores, will have what you need. More cities now have stores that specialize in home pest and weed control supplies. Your best source to find stores that carry equipment is the yellow pages of your telephone directory, usually under Pest Control Supplies or under Agricultural Supplies.

As they say, "Let your fingers do the walking." A little time on the phone, followed by visits to a few places, can pay off, not only in getting the best price for quality equipment but also in locating the place where you can get all the chemicals, repair parts, and even free advice you might need.

HAND TANKS

The most common piece of equipment is the pressurized hand tank, which is used for liquids. With most units, the pump is set into the tank and doubles as the handle. Air is pumped into the tank and the chemical is forced through the hose and out of the nozzle.

These units come in sizes from about ½ to 3 gallons. Also available are backpack units of higher capacity (usually 3 to 5 gallons), but these have limited value for the average homeowner. The 1-gallon size is the usual choice.

The standard of professional exterminators is made by the B&G Equipment Co. This is a stainless steel tank, with brass pump cylinder and fittings. It's pure quality throughout, and consequently, is much more expensive than anything else you're likely to find. Even

A large and small plastic spray tank.

their ½-gallon size will cost more than $150 in most areas, with the most common 1-gallon size approximately $210.

Another reliable "grandfather" company is Hudson Manufacturing. They make stainless steel, galvanized steel, and plastic tanks in a variety of sizes. Their strength is that they produce a line of products for the homeowner on a more limited budget.

With the growing trend of "do-it-yourself" pest control, a number of other companies have come about. Most concentrate almost exclusively on plastic tanks and relatively cheap parts. These tanks work reasonably well, but keep in mind that you get what you pay for. There is a reason that you can find a 2-gallon sprayer one place for $20, while another carries a price tag of more than $200 for a smaller, stainless steel tank with brass fittings.

Of the metal tanks, stainless steel is superior because of it's ability to resist corrosion. A fair second are metal tanks, which are lined with various substances (usually plastic) to prevent corrosion.

Galvanized steel is another common type of metal tank. It's major advantage is cost (it's cheaper) but it has a greater tendency to corrode. Some chemicals, such as Roundup, carry warnings on the label that they should not be put into galvanized tanks. Even so, these tanks can last for many years, as long as you remember to thoroughly clean the tank after each use. (For that matter, *no* tank will last for long without proper cleaning after each use.)

A good compromise of utility, weight, and cost is plastic. It tends to be relatively impervious to corrosion, is light in weight, and is less expensive to manufacture. Unfortunately, most of the "bargain" plastic tanks are made with the cheapest possible parts.

The smallest, least expensive, tanks have the nozzle built into the handle. (These are almost invariably the plastic tanks that hold only a quart or two.) That's fine for some jobs, but for general and more versatile use, get a tank with a hose attached to a wand.

The nozzle can be adjusted from stream to a fine spray. The least expensive tanks generally adjust in much the same way as a garden hose nozzle. The end is twisted, and the spray comes out in a cone shape. If you can afford one of the better tanks, get one with a nozzle that sprays out in a fan shape. This is more versatile, and safer.

One-gallon pressurized hand tank.

Stainless steel sprayer.

Small capacity stainless steel sprayer.

Cleaning and Repairing

Although it has been mentioned several times, proper care of the tank is critical, regardless of the material, and cannot be stressed enough. After each use be sure to clean the tank, hose, and nozzle with clean water. Empty the tank (preferably by using up the chemical rather than just dumping it, which is, in many cases, illegal). Fill and rinse the tank with clean water. Refill again with clean water, pump up some pressure and allow the water to spray through the hose and nozzle. Release pressure and empty the water that remains. Don't forget to drain the hose as well (by holding it up higher than the tank while there is no pressure and pressing the release).

The few extra minutes it takes to clean the sprayer will add many years to its life and ensure trouble-free operation. Make it a habit.

If the tank has been sitting unused for a while, it's a good idea to fill it with just water, pump up pressure, and spray. This way, if something has gone wrong with a hose, gasket, or connection, you won't find yourself being soaked with poison.

Repairing a hand tank is generally simple. The greatest problem is finding parts. If the local dealer who sold you the tank doesn't carry parts, you'll have to call other dealers and possibly have to write to the manufacturer. With the cheaper tanks, repair is often more expensive than buying a new tank. But don't throw out the old unit. Scavenge it for its parts. And the next time remember the advice given above, keep it clean, and never never store chemicals in it.

The first parts to go are usually the gaskets and hose. Most of these simply slide or pop into place. It's rare that you'll need more than a few wrenches, and possibly a screwdriver, to do the repair or part replacement.

If the sprayer has brass parts, be especially careful not to overtighten any threaded parts. Brass is quite soft, and you could easily strip the threads or damage the piece.

Should the nozzle clog, never use anything to clean it that is harder than the nozzle itself. A steel needle used on a brass or plastic nozzle will often make the hole larger than it was originally. After that, you might as well throw it away. It probably won't work properly.

Larger areas can be treated in a number of ways. Professionals most often use a portable power sprayer with a 50-gallon or larger reservoir, and a small gasoline engine to drive the pump. The chemical is pumped through a special high-pressure, chemical resistant hose.

This is an expensive method, however, and usually unsuitable for the average homeowner who might need this just once or twice per year. Even the smallest (about 15-gallons) can cost more than $600.

Most of the chemicals you will be using work just fine with one of the garden hose attachments. This attachment consists of a glass or plastic jar that serves as a reservoir for the concentrated chemical, and a special nozzle. The water flowing through the hose and nozzle creates a suction which draws up the chemical from the jar. It then mixes with the water and sprays out in the needed diluted form.

A few attachments have a trigger. Most, however, have a small hole which, when covered, creates the suction, and when uncovered, stops the suction so that only water sprays. There is usually some kind of set-screw or knob to adjust the dilution rate. This setting is rated either in tablespoons per gallon, or ounces per gallon. (Two tablespoons equal one ounce.)

HAND DUSTERS

Insecticide dust is used to treat plants, dry areas, and other specialized treatments. Trying to apply it by sprinkling it directly from the container is inefficient, wasteful, messy, and even dangerous. In almost all cases, if you can plainly see the applied dust, you've applied it too heavily. (The exception is when you're purposely applying a bead of dust, such as around an ant hill.)

In most cases, dust is best applied with a hand-pumped duster. These have a reservoir for the dust (usually about 1 quart or less). A plunger draws the dust from the reservoir and mixes it with air. The air then acts as a carrier, just as the water does with a liquid sprayer.

Small "garden" dusters can be purchased for just a few dollars and are fine if you intend to dust infrequently. They tend to be unreliable, however. The dust might come out in clumps, or not at

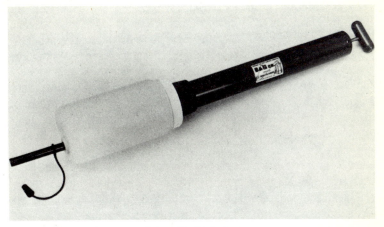

A hand-held duster.

all. Within a relatively short time the plunger may break, or the canister might develop leaks.

If you plan to dust often, buy one of the larger, better models.

Power dusters are also available, but are expensive. Fortunately, you won't often come across a need for one.

The simplest unit looks much like a large fire extinguisher with a removable top. The top is taken off and the tank is partially filled with dust. Then the top is put back in place and sealed tightly. Air is pumped in by an air compressor (like the hose at a gas station). Pressing the trigger allows the pressurized dust/air mixture inside to be sprayed out under force. It's messy, but effective.

Another type of duster uses an electric motor, much like a hand vacuum cleaner. Generally, these units spread the dust more finely, but they tend to be more expensive.

GRANULES

Granules are used in damp areas (outside). A special clay is impregnated with the active chemical. As the clay dissolves, the chemical is released. This makes it particularly good when treating areas that are constantly damp (not soaking wet!) or areas that might become damp. It also means that the moisture of your hand will also activate the chemical.

7

Never spread granules by bare hand.

A readily available device, manufactured by Ortho, is the Whirlybird. The cost is about $10, which is very reasonable considering the quality and versatility. (It can be used for spreading grass seed, fertilizer, granules, etc., and will last for many years under ordinary use.) A reservoir holds the substance to be spread. A knob on the side sets the size of the opening at the bottom of the reservoir, allowing the Whirlybird to be adjusted to the size of particles to be spread. A hand crank causes a bladed plate to spin and disperse the substance evenly.

Cleaning dusters and granule spreaders is usually difficult. Both dust and granules tend to clump when damp, which means that if you clean the device with water, you have to let it dry completely before using it again. Any dampness can cause it to clog. And because of the construction of most units, you probably won't be able to take it apart for better drying, or for repairs.

A WORD OF CAUTION

No matter what kind of equipment you're using, *never* mix chemicals in the same unit. If you have to change chemicals, thoroughly clean the unit and all parts first. Unless you're a qualified chemist, you won't know what the result will be. It could end up being totally useless, or extremely dangerous.

A common mistake is to use the same spray tank for both insecticide and weedicide. Not only do you run the risk of mixing chemicals, you could end up with a slight residue of the weed killer and end up destroying the plants you're trying to protect from insects.

Don't think that you can wash out the tank completely. Although the chemicals are water soluable in most cases, it takes some doing to remove every trace. (Would you want to drink from the tank?) Don't take the chance. Spend the little extra needed to buy a second and separate tank for weedicides. The cost is less than the cost of replacing a lawn or garden. (And don't forget to label the tanks appropriately! It doesn't do much good to have two tanks if you can't tell them apart.)

2

Chemicals

Professional exterminators are sometimes looked on with awe. One look in the chemical storage box and their arsenal of bottles filled with strange, evil-smelling liquids seems only to increase that impression.

Professional killers, handling deadly substances? Ask him what he's using.

"Uh, well, it's an organo-phosphate."

"An organo-whatsphate?" you ask.

"Organo-phosphate. It's based on phosphoric acid."

Great. That really explains a lot. It sounds vicious, like something out of a Vincent Price movie.

Actually, there is nothing mysterious, vicious, or evil about most insecticides. The nicotine in the cigarette you smoke is hundreds of times more dangerous than most insecticides.

Of course, all insecticides have the potential of danger. Sickness, even death, can occur if they are not handled properly. Respect them! But, with a little common sense, there is no need to fear them.

They are designed to kill insects weighing a fraction of an ounce. As long as you follow the directions printed on the label, they will do this and no more.

The first step towards safety is simple, but one that many people refuse to take. *Read the label!* It's easy to do, and will take very little time. The label tells you exactly what the chemical is, how to use it, mix it, and store it, where to use it, where *not* to use it, what it will kill, and even how to dispose of the container. Everything you need is right there. Government regulation requires proper labeling, for *your* protection and the protection of the environment.

The few minutes you invest by reading it will make the job more enjoyable, more effective, and *safe*. There are almost never cases of accidental poisoning when the label directions are followed.

People who manufacture chemicals have spent thousands, sometimes millions, of dollars on research. They know how to use their product to obtain the best results. All that time and money are put right there for you on the label. Make use of it.

Many people disregard the most basic rules of safety. Chemicals are stored in the open where children can easily get their curious hands on them. I've seen some stored under the kitchen sink, or even in the same cupboard where food is kept.

Chemicals should be stored in a well-ventilated, *locked* area. Even then, they should be placed high, and out of reach. I knew a lady who kept a container of chlordane in a locked room, but one day forgot to lock the door. Her three-year-old, while exploring, came across the poison (which was on the floor) and proceeded to give himself a bath.

Keep poisons out of reach! A little stretching is good exercise for you. It's much better than the exercise you may get when your child or pet has to be rushed to the hospital. Even if *you* don't have children or pets, your neighbor might. Remember, you are responsible for what happens on your property.

Another consideration is that light, moisture, and heat all affect most chemicals. Many come packaged in light-proof or light-resistant containers. That takes care of part of the problem. The rest is up to you. Store the chemicals properly, as stated on the label. Also use your own common sense.

A container can only keep out moisture and reduce evaporation if the lid is kept on tight. Cardboard or metal containers require special care because moisture can affect either, and thus affect the chemical inside. A cardboard box might collapse. A metal container might rust and corrode.

Glass can also be dangerous because it can break. You could end up with dangerously-sharp glass shards, possibly coated with deadly poison.

The dilute spray you apply around the home is relatively safe (which doesn't mean you should handle it carelessly). But to make a diluted spray, you will probably be using concentrated poison. A chemical that is diluted 1 ounce per gallon is 128 times more potent in its concentrated form.

You can buy premixed chemicals and avoid the problems of mixing your own, but this means you'll be paying a premium price for mostly water. For example, a chemical might call for a dilution rate of 2 ounces per gallon of water and cost $5 per quart of concentrate, resulting in 16 gallons of finished spray. Buying it premixed might cost $5 per gallon, or 16 times as much as the concentrate.

Handle concentrates with care. Wearing chemical-resistant rubber gloves is a good idea. Should you get the concentrate, or any chemical, on your skin, wash it off with soap and water *immediately*. Many pesticides can be absorbed through the skin.

To get rid of empty containers, follow the directions on the container. If you are in doubt, contact the nearest branch of the Environmental Protection Agency (E.P.A.) or Department of Agriculture. Proper disposal not only decreases the chance of a child accidentally poisoning himself, but helps to protect the environment. And, you'll be obeying the law.

Children have a fondness for using things they find in the trash. An empty bottle or can might seem to be a perfect "canteen." More than a few children have been poisoned this way.

In all cases, reusing the container is unwise. No matter how well you clean it out, there could be a trace of poison left. As stupid as it sounds, people have been known to store food and water in empty pesticide containers. *Don't!*

By the same token, don't use bottles to store chemicals which might be mistaken as food containers. Pop bottles are a favorite. The practice is stupid, and illegal. Even if you mark the bottle, remember that a great number of people don't bother to read signs, and many children can't.

There is no such thing as being too careful. Just when you think you've taken every precaution, take the time to figure a way to go one step further. Children and pets are ingenious snoops.

For measuring, get yourself an unbreakable plastic cup marked in ounces. One marked in partial ounces is even better. Don't use the kitchen measuring cup. (You'd be surprised how many people do.) Finally, after each use of the measuring device, wash it thoroughly, and lock it up with the rest of your equipment.

Washing out the cup is important for several reasons. The most obvious is for safety. Also, the next time you want to use it, it will be clean. If you happen to be using a different chemical, the two concentrates won't end up in the same tank.

Certain insecticides, when mixed together, cause a chemical reaction. The result could be a useless spray, or a spray that is potentially deadly. An unclean measuring cup can cause nothing but trouble.

Never mix insecticides together. Unless you're a qualified chemist, you can't know what the end product will be. You're much better off with the bugs. Mixing is also illegal. Even unintentional mixing can bring a $1,000 fine. The fine for intentional mixing can be as much as $5,000 and up to 30 days in prison.

The laws for pesticide misuse are quite strict, and with good reason. Enforcement of these laws is becoming more strict every day. Even if you are one of those who have no respect for the law, think of what you might be doing to the environment, or to your own child.

The various pesticides can be bought at a number of places. Most nurseries carry a variety of insecticides. "Do-it-yourself" stores often carry them, as do many drug and department stores.

Pesticides come in various formulations. Which one you want depends on what pests you have, and where they are.

Liquid concentrate is the most common form of pesticide. A few are designed to be mixed with oil. Although these have a better initial kill, they are much more difficult to handle, are more expensive, and have the unfortunate habit of staining. It's usually best to avoid them.

Water emulsifiable concentrates are better. Water is readily available, is easy to handle, and doesn't stain nearly as easily as oil.

The concentrate doesn't actually dissolve in the water, but breaks up (emulsifies) into millions of tiny droplets. Light refracts off these droplets, giving the finished spray a milky appearance. The water serves as a carrier. Once it evaporates only the chemical remains.

Wettable powders are just what the name implies. The insecticide is in the form of a powder, and as with concentrates, water is the carrier. Powders have the advantage of not ''weathering'' as quickly as the emulsified chemicals. But, they are difficult to work with and often clog the sprayer. Chances are, you'll never come across a pest problem where some other formulation won't work as well, or better.

Granules are usually made of a special clay, impregnated with the insecticide. As moisture dissolves the clay base, the chemical is released. In wet areas, they are more effective than liquids or dusts. In dry spots, they are almost useless.

Few things work better on ants than granules. Because of the ant's social nature, the ant will carry the granule right down into the nest. Once there, the relatively higher humidity activates the poison.

Granules are easily broadcast because of their semi-pellet form. Wind won't blow them away as it might dust. One caution: *Never* use bare hands to spread them. The moisture of your skin will activate the poison. A Whirlybird is an inexpensive device made specifically for spreading granules (see Chapter 1). If you plan to use them only on ants' nests, you can get by just sprinkling the granules out of the container. Don't overuse them, however. It doesn't take much.

Dusts are fine particles of either clay or talc that have been ''soaked'' in insecticide. They are best used in dry areas, and in spots where no other formulation will work. For example, the insulation in an attic will only absorb liquids, and granules would be useless because of the lack of moisture. Another spot to use dusts is in wall voids, and other closed-in places. A good duster that disperses a fine application is invaluable here. The tiniest air current will carry the dust further in.

Even in dead air spaces, the very process of pumping the dust will provide sufficient air current to move the dust around.

Again, don't over apply them. A good rule of thumb is that if you can see the dust, it's too heavy. There are rare exceptions, such as when you place a bead of dust along the crack outside a door, or when you put a ring of dust around an ant hill. Even then, a thin bead works as well as a heavy one. Better. And it is safer.

Dusts use air as a carrier. Basically, wherever air can go, dust can go. It sticks to walls, beams, and so forth. Like wettable powders, their life span is usually longer than liquids. But if there is excessive moisture present, dusts will only cake up, and become useless.

Plants absorb liquids quite readily. Dusts, then, are safer for treating plants. On edible plants, this means less chance of poisoning those who eat it. (Be sure to wash it off, of course.) On almost all plants it means less chance of burning.

If the ground isn't too wet, dusts can be used as tracking poisons. A circle placed around an ant hill will assure that all ants entering the nest carry a cargo of poison on their legs.

Because of its airborne nature, you'll have to be careful not to breathe in the dust. Your lungs are particularly susceptible to chemicals. Also, dusts cling to surfaces. Wash yourself thoroughly after using them, including your hair.

Aerosols need the same precautions as dusts. Being in the air, it is easy to inhale them: be careful.

The chemicals in aerosols are often oil based, which increases both the shelf life and toxicity. They are in cans pressurized with air or a gas (such as freon) as a propellant. Some are meant to be released only while you hold down the trigger; others are "total-release bombs" which empty themselves when the trigger is locked down.

Either way, the size of the nozzle opening is critical for effectiveness. If you can see the chemical fall, the nozzle opening is too large. A fine opening is much better unless you specifically need a heavier stream (such as when shooting a wasp nest). The resulting particles of a fine nozzle are smaller, and will drift farther and stay in the air longer.

A good aerosol is worth its weight in gold, especially for pests hiding in walls, cracks, or other inaccessible places. Only a few have a residual chemical, and most of these have a short life span. (You don't want one with a long life span because the chemical will be coating everything.) The primary function of an aerosol is to flush out pests so they will touch the residual chemical you've already applied, or to get a quick kill.

Most aerosols contain pyrethrum, with piperonyl butoxide as a synergist. (A synergist is a chemical added to increase the effectiveness of the main insecticide.) Neither chemical is considered dangerous, but they should be respected. They can cause an allergic reaction in some people. If you're one of those people, find something else to use.

Baits come in different forms, depending on which pest they are designed for. Snail and slug baits might be pellets or granules. Insect baits are usually powders or grains, substances the insect might be attracted to, thinking it's food. Rodent baits might be cracked grain, pellets, or powders which are dissolved in water.

Don't try to mix your own baits. They're touchy things. Besides being potentially dangerous, if the bait is too strong, or too weak, it won't work.

Commercially made baits work just fine. Except for snails and rodents, you'll probably never need to use a bait anyway.

TYPES OF INSECTICIDES

There are four major categories of insecticide: chlorinated hydrocarbons, organo-phosphates, carbamates, and botanicals.

Chlorinated hydrocarbons, such as chlordane, have pretty much disappeared from use. They have an extremely long life span, and a tendency to build up in the fatty tissues of the body. The most unfortunate action of chlorinated hydrocarbons is that they can be passed along the food chain. A bird or fish that eats poisoned insects takes the poison into its body. Each day it feeds on those insects increases the level of poison in its own system. Likewise, if another animal feeds on the bird or fish, it inherits the increased dosage.

It is believed that chlorinated hydrocarbons work on the central nervous system. In larger animals, accidental poisoning shows itself in several ways. Nausea and diarrhea set in, along with damage to the liver and other organs. Severe poisoning might result in convulsions, unconsciousness, or death.

Once chlorinated hydrocarbons were found to be dangerous both to man and the environment, scientists turned to a second group of insecticides called the organo-phosphates. These toxins are based around the phosphoric acid molecule, and usually have a short life span when compared with the first group.

They enter the body through the lungs, or through the stomach. Although the rate of absorption is slow, they will also go into the skin.

Poisoning by carbamates is caused by a reduction of the cholinesterase. Symptoms of poisoning are excessive salivation and sweating, constriction of the pupils, and loss of coordination.

Botanicals are derived from plants. Pyrethrum, the best known botanical, is extracted from a variety of chrysanthemum. Botanicals tend to have extremely short lives, and tend to be the safest of the insecticides. There has never been a death traced to pyrethrum. The most common problem is that some people might react allergically.

Should accidental poisoning occur with any chemical, get help *immediately*. Unless the victim is in dangerous surroundings, don't waste time trying to move him. Call a doctor as soon as possible, and give him all the information you can (chemical used, etc.). Wash off the chemical. The longer you take to do this, the more poison will be absorbed. Act quickly and calmly.

Better yet, be careful in the first place. The chemicals won't be at all dangerous if you use them correctly: read the label.

Baygon. Baygon is a carbamate, with an odor that has always reminded me of mothballs. It has an extraordinary "knock down" power against a number of insects.

The two most common forms are liquid concentrate and bait (for roaches). The liquid is fairly dilute to begin with, requiring 4 to 8 ounces of concentrate per gallon of water. The bait is already prepared at the proper strength (usually 2 percent).

Baygon has two unfortunate habits. First, it is known to stain or pit many materials. Rugs, all kinds of cloth, and some rubber, plastic, and asphalt materials may be damaged. Second, it tends to crystallize, especially in cold weather. There is nothing quite so frustrating as having to completely tear down the sprayer because these crystals have jammed it up.

So, stay away from it unless you know how to tear down (and rebuild) your sprayer. It may not happen, but if it does, and you don't know how the sprayer goes together, you could find yourself in a bit of trouble.

When used around porous materials, like unpainted wood, baygon is all but useless. It requires a fairly solid surface. Always be careful where you apply it. It is toxic to smaller life, especially birds and fish. If there is any chance of birds landing on it, or of the chemical washing into a lake, pond, or stream, don't use it.

Boric acid. Boric acid acts as a desiccant, which means that it destroys the waxy coating of the insect. The insect then has no protection from the drying sun. Within a very short time it dies of dehydration.

Most often, boric acid is purchased as a powdery crystal. A light application along doorways, and in cracks where the pests crawl and hide works wonders. Boric acid has a great advantage in that it isn't actually a poison.

Carbaryl. Carbaryl is better known as its brand name, Sevin. The two names are used interchangeably. It belongs to the class of carbamates. It has replaced DDT and parathion as a crop dust. Carbaryl is often used in flea and tick powders. It is not only very effective against these pests, but is also safe to use around the infested animals.

As mentioned before, carbamates are cholinesterase inhibitors. However, the action is easily reversed. This same action when caused by the organo-phosphates, tends to be irreversible. This makes carbaryl a generally safer chemical.

Carbaryl is made as a liquid concentrate, dust, or as granules. Sometimes it is used in conjunction with pyrethrum or rotenone. Its life span is fairly long, but it does not build up in the body, or in the environment.

It has a particularly deadly effect on bees. But, before they die, the bees tend to go ''beserk.'' After being hit with carbaryl, they'll sting anything. If you have plants that need bees for pollination, use something else.

Certain plants can also be burned. Read the label before you use it.

17

Chlordane. From its name you can tell right off which chemical group this belongs in. It is a chlorinated hydrocarbon, perhaps one of the best known (next to DDT).

As of July 1975, the E.P.A. declared the used of chlordane, except for application for subterranean termites, to be illegal. As of April 15, 1988, its manufacture, sale, and use, other than of stock on-hand by a homeowner, is strictly illegal. Once the present stock is gone, if you have any, there will be no more. The reasons given were the long life span, the tendency of it to build up through the food chain, and the possibility that it *might* cause cancer.

The liquid concentrate is good for many insects both inside and out. Since it has a fuming action it is excellent against spiders, scorpions, and centipedes. It is highly effective against ants. (Dusts or granules are better than liquid on ants.)

Liquid chlordane has a higher potential of staining than does almost any other insecticide. Even professional exterminators didn't like using it inside, or any place where staining could be a problem.

Because of its ability to stay in the body, building up when used often, caution must be observed. If you are using it as a liquid or dust, be careful not to breathe it. At best, you'll be coughing and sneezing for the rest of the day. And, whatever your body has absorbed that day will still be with you for years to come.

DDT. The manufacture, sale, and use of DDT is *strictly* illegal. Present laws won't even allow for its disposal. DDT has an extremely long life, and can be a hazard to the environment, and to life. If you have any, *don't use it!* Contact the local E.P.A. for instructions as to what to do with it.

Don't dispose of it without their telling you how to go about it. Chances are, they'll just pick it up. Throwing it in the trash can is illegal, and dangerous.

The laws that surround DDT represent a perfect example of what happens when people abuse an insecticide. Its popularity was such that it was grossly over-used. Had it been used with a degree of intelligence, it wouldn't have presented a danger. But, it was sprayed and dumped indiscriminately, and the result was the death of millions of birds, fish, other wildlife, and even a few people.

DDVP. Chances are you will never come across this chemical unless it is premixed as a "booster" with another insecticide. Even then, the E.P.A. is considering restricting or banning the chemical because of its harmful side effects on small animals. Their concern is especially strong when it comes to using the chemical in any confined area. It is readily absorbed by the skin and through the lungs.

Its chief action is as a vapor, which gives it its second name of Vapona, a brand name owned by Shell Oil. For some years they used it in their Shell Pest Strips and also sold it as a liquid concentrate.

It works primarily as a flushing agent or as a quick, direct kill, although it also has some residual action. The fumes reach into cracks and crevices. If the pests aren't chased out across the previously (or simultaneously) applied residual chemical, DDVP can kill them where they are.

It's best to avoid using DDVP, especially if there are children or pets around. Using it in a room where there is a fish tank can be disastrous. You'd possibly end up with no bugs, but also no fish. In short, even if the E.P.A. allows the continued use of DDVP, it's not a chemical for the homeowner.

Diazinon. At present, diazinon is one of the more effective legal chemicals. It is an organo-phosphate, and can be used as a liquid, dust, granule, and occasionally as a bait. A common brand name is Spectracide, manufactured by Ciba-Geigy Corporation.

It is slightly more toxic to man than malathion, and is absorbed through the skin, lungs, or the stomach if swallowed. Handle it with care and respect.

It can also damage certain types of fern, hibiscus, and gardenia (among other plants). Be sure to read the label before using it on any plant. If in doubt, write directly to the manufacturer for further information.

Diazinon is particularly toxic to fish, birds, and other small wildlife. Apply it carefully, and don't let any wash into streams or lakes. Small children may also be susceptible.

Despite this, diazinon is probably one of the best insecticides you can buy. If you use it as directed on the label, you'll have no problems.

Heptachlor. Along with its close relative chlordane, heptachlor is going into a gradual oblivion. Both substances were declared dangerous, and manufacture is now illegal.

By far, the most common form of heptachlor was in granules. Few things can equal its effectiveness against ants. A light sprinkling of granules around an ant hill was sufficient to completely eliminate the problem.

The other major use of heptachlor was as a "booster." When mixed with chlordane (liquid), the two became an effective, and long lasting, treatment for termites.

The precautions that apply to chlordane also apply to heptachlor. It has an extremely long life span, and builds up in the body. Repeated use can be dangerous to you, and to the environment. Even if you happen to have some heptachlor lying around, try using the safer chemicals first. If you do use it, do so sparingly.

Kelthane. Kelthane is a chlorinated hydrocarbon, closely related to DDT. Its actions are basically the same, and it should be treated as a dangerous substance. Avoid it.

Lindane. Lindane is an extremely dangerous chlorinated hydrocarbon. It has been used on cotton insects and termites as well as many other pests. The vapor action of lindane makes it effective against flies, spiders, and mosquitoes. Sometimes it is manufactured in a tablet form, which is heated to give off the fumes.

The use of DDVP pretty much replaced lindane. It is much safer, and nearly as effective in most cases. Lindane has caused several deaths, and is particularly toxic to calves.

Although more toxic than DDT, lindane doesn't build up in the body to such a degree, nor does it have such a long life. Still, it is dangerous. Avoid it whenever possible.

Malathion. With an odor somewhat similar to boiled cabbage, malathion is perhaps on the most widely used of all organophosphates. It absorbs through the lungs, or through the stomach if swallowed. To a lesser extent, it might be absorbed through the skin.

It can be obtained both as a liquid concentrate, and as a dust. Occasionally it can be used as the active ingredient in certain baits, or even as a wettable powder.

Malathion can kill fish, and is also toxic to many kinds of birds. (Mixed with grease and spread on roosting spots, malathion has been used to kill pest birds.) Be careful when using it around these animals, and never let it run into any body of water. Mother Nature will love you for your consideration.

The life span might be anything from a few days to two months, depending on conditions. It is broken down quickly by heat or sunlight. While this is a definite plus where the environment is concerned, it also means that you must be certain to store it properly. Keep it out of direct sunlight. The dark-tinted bottle will help, but you must still take other precautions.

Along with being safer than most other residuals, malathion also has less tendency to stain. This doesn't mean that it won't, so apply it carefully.

Methoxychlor. Methoxychlor is one of the few chlorinated hydrocarbons still legal. It is safer to use than any other chemical of the same group. In fact, it is even safer than malathion.

About the only way you'll ever see it is as a dust, usually under the title of "Rose Dust." Occasionally it is mixed (*by the manufacturer*) with malathion in a liquid concentrate.

Unlike the other chlorinated hydrocarbons, methoxychlor has a short life span, and so provides little threat to the environment. Use it as directed on the label, and it can be safe, and effective. Its major use is in the garden, against such pests as aphids.

Parathion. Until recently, parathion was used as a crop dust. Something I'll never understand. It's extremely toxic, and quickly gets into the body through the skin, lungs, or by ingesting it. Poisoning occurs rapidly, and serious illness or death is not uncommon.

There are several instances where parathion leaked into the cockpit of the crop dusting plane. Almost invariably the pilot passed out, and crashed. In one case, six children were watching a crop duster, and had some of the dust drift across them. Of the six, three died, one was permanently blinded, and another has permanent nerve damage. Only one came out of the experience unscathed.

Stay away from it! Even if you happen to come by some, which you shouldn't be able to do, don't use it—anywhere!

Phosphorous. Phosphorous, usually in the form of a white, yellow, or red paste, has been used as a rodenticide. It is meant *only* for those who know exactly how to handle it. *Stay away from it!*

Poisoning occurs mainly through ingestion. But, there was a case where a three-year-old walked barefoot on some that had been applied to a baseboard. The child died. At best, skin contact might cause serious burns.

Pyrethrum. Along with the closely related Alrethrin, pyrethrum is a botanical. (Another botanical is strychnine.) The most common commercial pyrethrum is actually two types of pyrethrum, both extracted from plants.

It has never been known to cause a death, even when it was swallowed in a suicide attempt. Its life span is extremely short after exposure to air or light. Consequently, the chances of it harming the environment are nil. The only problem with pyrethrum is that it can cause an allergic reaction in some people.

Most often it is found as an aerosol (about 1 percent). Another use for pyrethrum is as a flushing dust. Insects touched by this dust are chased from their hiding places, and across the residual. Sometimes a residual, like diazinon, and pyrethrum are mixed together in a dust.

With the outlawing of chlordane, even for use against subterranean termites, synthetic pyrethrum has been used for this application. The most common brand name of this synthetic blend used for this is *Torpedo*, possibly named because it is applied beneath a structure, where the effects of air, sun, and heat won't affect it.

Its entry is made through the skin. Insects die rapidly when they contact the chemical. In an aerosol, it may take just a few tiny droplets to kill a fly. The fly seems to "go crazy" and can be seen flying into objects, particularly windows. Later, you'll find the dried-up carcass on the window sill.

Rotenone. Rotenone is very similar to pyrethrum, and is used in much the same way. Both are botanicals, and are safe to use around warm-blooded animals. However, rotenone tends to be toxic to fish.

Because it is extremely difficult to dissolve, retenone is usually manufactured as a dust. Occasionally it may be found in aerosols.

Silica. Several "insecticides" rely on silica (sometimes called silica aerogel) to kill. A common brand name is Permaguard. Like boric acid, silica works as a desiccant. The tiny crystals wear away the protective coating of the insect. The result is the dehydration of the body fluids.

Although it can be classed as an insecticide, silica is *not* a poison. This makes it extremely safe. If you breathe in the powder it might have a drying effect in your lungs, but won't actually do any harm.

Warfarin. The use of warfarin and pival (a type of warfarin) have replaced the use of phosphorous, thallium, and other dangerous rodenticides. When used as prepared by the manufacturer, it is safe.

The recommended strength is .025 percent. Although this seems to be a small amount, tests have shown it to be the most effective while still being safe. At this strength, larger animals are less likely to be harmed should they accidentally eat some of the bait. It takes approximately the same amount of bait as body weight to kill. A five-pound dog would have to eat nearly five pounds of the bait to kill him.

That doesn't mean you can be careless with it. It is designed to work on warm-blooded animals. Even if it doesn't kill, the internal problems can be unpleasant.

It works as a stomach poison, and does not absorb through the skin in any appreciable amounts. The poisoning takes place in two ways. First, it inhibits clotting of the blood. At the same time it causes internal hemorrhaging.

The rodent virtually bleeds to death. Because of the dehydration effect, the rodent will try to get to water. If none is available inside the home, it will go outside.

The ulcerating of the stomach causes nausea. The animal will seek blades of grass to induce vomiting, which is another force that drives him outside. Even if he finds the grass, a mouse cannot vomit. He merely ends up choking to death.

Besides the relative safety of warfarin, it has yet another advantage. Unlike the other more dangerous rodenticides, warfarin has less tendency to have a double action. A cat that eats a mouse killed by warfarin stands less chance of being poisoned himself. The only poison that affects the cat is what happens to remain in the stomach of the mouse. Accidental poisoning *is* possible, however. So, be sure that the family cat doesn't get at the dead rodent. And, place the bait out of sight and reach.

Fumarin is one more type of warfarin. Usually it comes in a soluble powder. Mixed with water, and perhaps a bit of sugar, the rodent will not only be eating the poison, but also drinking it.

Of course, baits are all but useless if there is an abundance of other food sources around. The same applies with the liquid baits. Before you place the baits, remove all other sources of food and water.

3

Home Maintenance

The best method of pest control is *not* the application of poisons, any more than the best method of medical care is swallowing pills. The key is prevention—having conditions where the problem doesn't occur, or at least doesn't get worse, in the first place.

Proper home maintenance and cleanliness is the first step towards effective pest control. The best exterminator and the deadliest poisons won't be able to control pests under unsanitary conditions.

Start by giving your home a complete survey. Look at the surroundings. Is the yard cluttered, or overgrown with plants, grass, and weeds? Are there potential nests around your house? How about sources of food and water?

Does your house have cracks or holes where the pests can gain entry? Are the screens tight and in good repair?

Is the inside clean? This means behind and under objects as well as the counter tops. If grease or dropped food can accumulate behind the stove, you can bet that you've just invited all sorts of pests to dinner.

Most pests come in from the outside. This is probably where you'll want to start your repairs. If you live in an area that has dangerous pests (scorpions, black widow spiders, poisonous snakes, etc.) be extra careful. Use a stick to turn over boxes, boards, and rocks. It might also be wise to wear heavy gloves.

Cluttered, unkept areas are natural hiding places for pests.

Should you feel something crawling on you, *never* slap. Brush. Slapping can drive a stinger into the skin. By brushing, you'll push the creature off, and reduce the chance of injury.

Remove all debris from the yard. Stack wood neatly, and at a distance from the house. If at all possible, elevate it. Even then, you'll have to pay special attention to this area. It is an open house for many pests. Check it regularly.

Weeds should be torn out, and the grass kept clipped short. Dead and decaying vegetation must be removed. A bag to catch lawn clippings is a cheap investment.

Plants close to the house are particularly bad, especially when they are neglected. A rake is wonderful for keeping dropped leaves away. Use it regularly.

There isn't much you can do about the crack where the ground meets the house. Many insects love this spot. But if the area around it is kept clean, they won't be quite so enthusiastic about living there.

Leaky faucets should be fixed. Usually it isn't difficult. A washer will cost only a few cents, if you don't have one lying around already.

A wrench and screwdriver, and about 10 minutes, is all it takes. Even if the entire faucet needs replacing, it's worth the effort. Changing one is easy. Not only will you be saving water, but you will also be taking away an essential supply of water for insects and rodents.

Holes at all heights should be repaired. A box of patching plaster, or a tube of caulking will do the trick nicely in most cases. Larger holes may need a piece of wood or metal. You'll eliminate many nesting areas, and cut your heating and cooling costs. This well-invested time will also prevent pests from coming into your house.

Windows and doors must be sealed for the same reason. A scorpion can crawl through a crack of only $\frac{1}{16}$ inch. Many pests can creep through even smaller holes. The best chemicals take time to kill. In the meantime, the pest can make a considerable nuisance of itself.

Screens with holes or breaks need to be repaired or replaced. They are designed to allow you to let in air without letting in the bugs. If they have openings, they are useless.

Attic vents usually have screens across them. These should be checked occasionally. If there aren't any screens on yours, put some on. Screening isn't all that expensive. And, chemically treating the vent just won't work.

Anti-bug lights will help decrease the number of insects at night. Don't expect miracles from them, but they will keep away about 80 percent of the flying insects that come to a normal light.

Garage doors are famous for having huge cracks around them. If you try to seal up the openings, be sure that you aren't interfering with the operation of the door.

Inside the home, the process is basically the same. Look under the sinks where the pipes come in. Most have holes around them like the Grand Canyon. If you have a gas stove you may have to move it out (*carefully!*) to check the gas pipe. If there is a hole around it, repair it. Around all pipes, don't just assume that the metal ring means that the pipe is sealed. The primary function of that ring is to hide a hole.

And while the stove is out, clean behind it. Wash down the wall with a good cleaner to cut the built-up grease.

Cupboards may also have cracks where they meet the wall. Caulking compound will seal these quite well, and eliminate one of the favorite homes of roaches. If they don't have a place to nest, they won't be able to infest your home.

The caulking won't show in most cases. For those spots where it does, use a caulking that accepts paint. A small brush, and a few minutes, and you have what looks like a professional job.

Leaking faucets can cause damage as well as provide a source of water. They will have to be repaired in any case. The counter top may also have come loose from the water. After the wood underneath has been given a chance to dry, a good linoleum glue will tighten it up. Then, caulk the edges if there are any openings.

The area under sinks has a tendency to be a "catch all" for whatever doesn't fit elsewhere. At least keep it clean and dry. If you keep the trash there, use a plastic container, preferably with a tight fitting lid. Wet trash won't leak out, and water (and insects) can't get in.

The rag that many people drape over the water pipe should be clean, and as dry as possible. Better yet, avoid the habit. If left there for long periods of time, bacteria will grow, providing a delicatessen for insects. (You'd also be wiping your counters with an unsanitary cloth.)

Don't leave dirty dishes around. This is especially important if you already have a roach problem. Wash them immediately, and thoroughly. Badly washed dishes are almost as bad as unwashed dishes.

Spilled food should be cleaned up immediately. It's easy to ignore it, saying you'll get it later. Don't do it. The same for counter tops. Keep them wiped clean. If grease is a problem, use a detergent first, then use clean water, and then dry the counter.

Basically, this all breaks down to cleanliness and good sanitation. Even without insect problems, the reduction of bacteria, and possible food poisoning, makes it worthwhile. It might seem tedious, even phobic, but in a very short time it will become a habit. If kept up, the kitchen will always be clean. The time you spend there cooking and eating meals will be more pleasant.

Any place where water comes into the house will have many of the same problems as the kitchen. Check around pipes for holes. Lint might gather in the laundry room, and make excellent nesting material for many insects, and for mice.

The vent for the dryer is another spot where insects might come in. Be sure it fits tight. Any cracks should be sealed. If you put a screen over the vent, make sure it can be removed for cleaning. Lint will probably gather there, and unless you can take it off to clean it, it will cut the efficiency of the dryer.

Basements require special attention. They are often dark and damp, and make good homes for pests. Keep yours as clean as possible. Try not to let it become a cluttered mess. Water leaks must be fixed. Sometimes water will leak through the walls. A building supply store can tell you what to use on yours to stop this.

Attics don't usually have water problems unless there is a leak in the roof. (If so, repair it.) But, they are often left alone for years on end. If you seal any holes, you'll keep the insects out, and the pest problems to a minimum.

Finally, spend a little time in the crawl space, if you have one. Repair any holes or cracks. For mobile homes, duct tape will seal most of them, if they aren't too large. Check where pipes and ducts enter the house, and seal them. (You might also want to screen over where these vent inside the house.)

No matter how well you seal your house, insects always seem able to find a way in. But for each crack or hole you seal, that's one less chance they'll have.

The same rule applies to cleanliness. It might not completely prevent insects from finding food, but it will cut down the source considerably. Besides, good sanitation is its own reward.

Throughout your cleaning and repairing, remember what you are trying to do, and you'll know more exactly *what* to do. Eliminate the four things that pests need: food, water, nesting, and entrance.

4

Exterminating Procedures

The actual procedure of exterminating is based mostly on common sense. Unfortunately, there are times when even supposedly "trained" professionals fail to use common sense (often because their customers don't let them, but too often because of improper training).

For example, if there are no pests on the lawn, it doesn't make any sense to spray the entire yard. It's dramatic, and often used by pest control companies (at least in some areas) as a selling point, but is essentially useless as a routine procedure. The heat and sunlight will cause even a residual chemical to break down in a very short time. (One of the things the E.P.A. looks for when evaluating a chemical for public use is its tendency to break down. Chemicals that last for long periods of time under general conditions tend to be a threat to the environment, and hence are either restricted or illegal.)

It's a waste of time and money to apply a chemical where it's not needed. It also increases the potential danger to you, your family, and the environment.

At all times, the goal is to put down as little chemical as possible, especially inside. There are several reasons for this. Most important is safety. The chemicals available to the general public tend to be relatively safe. That doesn't mean, however, that they are completely safe. The less you use, the safer you are.

There are times when spraying the entire lawn *is* necessary, and also times when "fogging" the house—or a part of the house—is the only viable solution to a problem. These are normally special occasions, not routine. Most of the time you can get by with less. Sometimes with no chemical at all.

Your first priority is to remove or reduce the four needed basics—food, water, shelter, and entrance. The order depends on the pest being controlled, and the circumstances. (For a pest that stays outside, "entrance" will usually be either unimportant or impossible. A gopher has a "shelter" and a source of "food" that is generally outside your ability to control.)

Chemicals are an addition to, rather than a substitute for, other measures. These other steps are always your first priority. Only then should you begin to apply a chemical, once again, keeping the four basics in mind.

Insecticides are placed in, on, or near the nesting sites, and along the pathways between these sites and where the insect pest is getting its food, and where the pest is gaining entrance to the house.

Baits are placed in many spots. One of the primary functions of a bait is to replace the other source(s) of food and water. If the pest has other sources more readily available, your bait is likely to go untouched. Place the bait as near to the "shelter" as possible, and along regularly used pathways.

USING A SPRAY TANK

The "stock and standard" of pest control is the spray tank. A one-gallon tank is small enough to carry for fairly long periods of time, and can hold more than enough chemical to treat your entire home, inside and outside.

Remember that the idea is to put the chemical only where it is needed, not to splatter it about until the house is dripping. And you'll need very little, if any, chemical inside the home. (See A Standard Treatment later in this chapter.)

Most tanks have several nozzle settings, from stream, to a fine spray. Quality spray tanks have nozzles that spray in a fan shape. Less expensive units most often spray in a cone shape (like a standard garden hose nozzle). The fan shaped spray is easier to control since it can be turned to cover a wide area, or turned to cover almost no area. A cone shape always covers a wide area. About the only way to get good control, such as for treating the baseboards inside the home, is to adjust the nozzle to a tight cone instead of a wide one. Keep in mind that the resulting spray tends to be larger drops.

Many people have the tendency to overpump the tank. This isn't necessary, unless you need the exercise. It will also increase the chances of something going wrong with the tank, hose, nozzle, and connectors.

Lower pressure makes it easier to control the spray, which is why you should never overpump the tank for spraying inside or where splashes can be a problem. Outside it's less important, but even there a good medium pressure is best.

Most liquid chemical is applied outside, with the heaviest concentration usually going into the crack between the house and the ground. Anywhere else you spray depends on the particular pest problem.

For example, you might need to spray the eves for spiders. Do so very carefully, and keep in mind that the spray is going to drift down. Always spray away from yourself—never straight overhead.

Concentrate the insecticide in the cracks, where most pests hide.

If there is a breeze, let it work for you. Stand upwind and let that breeze carry the falling droplets away from you. (Do not spray in a heavy wind.)

There are legal restrictions on the amount of chemical that can be used inside. You'll never violate these regulations if you use your common sense. Confine your spraying to the baseboard where people and pets won't be walking, and in front of the doors if the pests are coming in beneath the doors. Don't spray indiscriminately. Have a reason for spraying a particular spot inside. And *never* soak down the entire floor.

Spraying Large Areas

There is rarely a need to treat large areas with insecticide. There should be a specific reason for doing so, such as a pest that is destroying the entire lawn.

The job can be done with a hand sprayer. If your yard is small, this is probably the preferred method since it allows maximum control.

Professionals usually use a power sprayer for treating large areas. A small gasoline engine or electric motor drives a pump which pulls the pesticide from a large tank (usually 15 to 100 gallons) and sprays it under fairly high pressure through a special chemical-resistant hose.

These units are expensive. Fortunately, there is an alternate solution for the homeowner—a hose-end sprayer (described in Chapter 1). These cost $5 to $15, depending on the quality, type, and where you buy it. The one you need should adjust for various dilution rates.

Most often the measurements are given in teaspoons or in tablespoons per gallon, which is strange because most of the insecticides you'll be using give dilution rates in ounces per gallon. Still, this is no problem. Just remember:

3 teaspoons = 2 tablespoons
2 tablespoons = 1 ounce

So, for a dilution rate of 2 ounces per gallon, the hose-end sprayer should be set for 4 tablespoons (or 12 teaspoons) per gallon.

A hose-end sprayer is an inexpensive and effective way to chemically treat large areas.

Read the pesticide label carefully before starting. Some chemicals have different dilution rates for larger areas than for smaller areas.

The yard has to be picked up first, This sounds obvious, but too many people don't bother to take the time and end up spraying the children's toys or the pet dog. No one, including pets, should be allowed on the lawn for a minimum of several hours, and preferably not for several days.

The day you pick for the job should be still. A light breeze probably won't hurt, but trying to spray in a wind is foolish. Safety is always important. If the wind can blow chemical all over your neighbors, and over you, put off the treatment for another day.

DUSTING

Dust tends to have a longer life span than liquid sprays. It's useless in damp areas, but is the preferred treatment in many others.

Treating plants is one example of when dust is often preferred. It has a much lower tendency to burn the plant, and it isn't absorbed

Spray on calm days only. Always aim downwards.

into the plant as readily as liquids. If the plant is edible, be sure to wash it thoroughly, regardless of the kind of treatment used.

Other places where dust is often superior are attics and crawl spaces. These places are usually dry, which allows the dust to stay active for very long periods. And since it is a solid (rather than a liquid), it won't be absorbed into the wood and insulation.

Dust is carried by air. This makes it possible for the dust to drift around and coat almost everything.

Consequently, be careful where you apply the dust. (You won't want to use dust where there will be people crawling around.)

Treating a large area, such as an attic, with a hand-pumped duster is difficult. Unless there is a natural air flow in the attic, you'll have to provide it. Although not much is needed, it will be hard to pump enough to send the dust more than 10 or 20 feet. However, crawling into the attic and dusting as you back out isn't always safe.

There is always the physical danger of falling through the ceiling. Worse, is that no matter how careful you are, while the dust is coating

Dust can often reach places liquids cannot, and is generally safer on plants than liquids.

the area to be treated, it's also coating you—including the inside of your lungs.

In other words, if you're going to be dusting such an area, do so with extreme caution. If you don't feel that you can do the job safely, don't even try. You can always resort to an almost equally good method of treatment - that of using aerosol bombs. (See the section, Aerosols And "Bombs," later in the chapter.)

A similar use involves smaller areas, such as wall voids (the gaps between the walls of your home). Here, even a small hand duster will supply enough air flow for the dust to reach deep inside. The problem is gaining access to that void. You'll probably have to punch a hole in the wall large enough to admit the nozzle of the duster.

A second use of dust is as a tracking powder. This means simply that the dust is put down so that the pest walks in it and then carries it back to its nest.

One of the most common pests to control this way is the ant. A small bead of dust is placed in a ring around the ant hill, several inches from the opening. As the worker ants go in and out, they carry some of the dust deeper inside. The goal is to have it affect the queen. (Unless you kill the queen, you won't have much luck getting rid of the nest.)

A bead of dust in cracks and in front of doorways can be very effective.

Dust can also be used, *carefully*, along the outside of a door. It's advantage here is its long life span.

Normally, dust is used only outside. It should never be used where people will be coming into contact with the dust. Keep two things in mind whenever you use dust: one, is that it has a long life span, and if kept dry, it is likely to still be effective months after it has been applied; second, is its nature of being carried by air. A small puff in a cupboard may not seem dangerous at first, but remember that the dust is going to drift and coat the entire cupboard with poison.

AEROSOLS AND "BOMBS"

Too often these aerosols and "bombs" are thought to be the first solution to a problem. What you have to keep in mind is that the insecticide in the aerosol is almost invariably dissolved in oil, which means that it will stain. On top of this, is the problem of polluting the environment with the gases used inside the aerosol. Even if you're unconcerned with staining, and unconcerned with the environment, the insecticides contained in most aerosols have a limited effect. What you spray out might kill off nearly 100 percent of what the spray hits, but an hour later there is nothing left.

There are two basic proper uses for an aerosol. The first and most common is for a quick kill. The second is as a flushing agent to drive pests from their hiding places so that both the knock-down effect of the aerosol and the lasting kill of the residual already laid down can do the job.

Unfortunately, too many commercially available aerosols don't do either job effectively. Many are designed strictly for one-time use, with a spray that is so heavy (droplets are too large mostly because the nozzle opening is too large) that it falls out of the air and becomes useless in a matter of seconds.

An aerosol that sprays in a stream is designed primarily to hit a target at some distance. The main use of this kind of aerosol is to kill a wasp.

Other than that, the aerosol you choose should generally have a small nozzle opening and fine spray particles. This is particularly important if the aerosol is to be used as a flushing agent (to get the bugs out of their hiding places) and even more important if the aerosol is to serve as a "bomb" (such as filling an entire room or house, with a "fog").

Fogging a house is *not* a standard procedure. If you do it, there should be a specific reason. This is because preparation for a proper fogging is time consuming if it's to be safe.

The job begins with a thorough application of a good residual spray, such as diazinon or malathion. Then it gets complicated.

All plants and animals have to be removed from the house, including the fish in your aquarium. (Don't forget to take out, or at least "bag," the tank. If you don't, the insecticide will be absorbed by the water.) It's also necessary to remove all open food, such as bread. (Canned goods, if they're still sealed, can stay.) It's also best to remove, or at least cover, any dishes or other eating utensils. The refrigerator can be used to hold some of the things *if* the door seal is good.

Next, you have to go around the house and shut off any pilot lights. Preferably, even the electricity should be shut off so that there

Aerosols are best used as a chasing agent to force pests out of their hiding places, or as a quick kill.

is absolutely no chance of a spark. Most "foggers" are not highly flammable, but many *can* ignite if given the chance. It doesn't make much sense to have your house explode or burn down just to get rid of some bugs.

A total release fogger.

Obviously, the entire house has to be closed up, and any ventilation shut off. If the windows or doors don't seal, or if there are holes in the walls, the fumes will be less effective.

Some bombs have the unfortunate tendency to leak, especially as the can nears empty. For this reason, several thicknesses of paper, preferably with plastic beneath, should be placed beneath the can to prevent any possible damage. It's also best to have the can slightly elevated, and as close to the center of the room as possible. Have this place—or places—ready before you begin.

Go to the spot farthest from where the can will be placed, or to the place with the heaviest infestation, and lock down the trigger, letting the bomb spray out and away from you. Work your way quickly to where the can is to be set—and get out fast. The time between when you first lock the trigger, and when you walk out the door, should be fast enough that you never breathe the fumes.

Then you have to find a place to go for six or more hours. The longer the better. For safety's sake, it's best to have someone come back early and open the house to clear any remaining fumes.

Fogging an attic or crawlspace is often just as complicated. First, there is the problem that the fog might find its way into the house. Second, there's always the chance that whatever pest you're trying to eliminate might find its way into the house. The reverse is also true—that fogging the house might chase the pests into the attic.

You do not always have to fog the entire house, attic to ground. That depends on the pest and the extent of the infestation. Moreover, it might not even be legal to do the job the way it needs to be done. For example, if your home connects with a neighbor's, fogging just yours will have little effect, and the fumes could seep into the other house causing problems for which you could later be held liable.

BAITS

Although there are baits for insects, they are usually used to control a rodent problem. For this purpose, baits can be highly effective—or almost totally useless. The use of rodent baits requires intelligent placement, and usually a lot of patience.

Keep in mind the four basic factors of food, water, shelter, and entrance, and the importance of knowing the habits of the pest. Placing bait without thought means that its effect will be a matter of luck.

The best bait, put in the best places, won't have much effect if the pest has other sources of food. Your goal is to remove as many of these as possible, to make the pest turn to the bait more often.

Where does the pest live? What paths are followed between its nest and its source of food? The answers to these questions will tell you where to place the bait. (Of course, always keep in mind that safe placement is always of first importance. The bait should be inaccessible to children and pets, and completely out of sight.) It should be placed where the pest will come across the bait first, rather than by accident.

A STANDARD TREATMENT

Most pests come in from the outside. This is where the routine pest control job begins. Quite often it's all you need. If you do a proper job outside, you might not need to put down any chemical inside.

The most important places to spray are where pests tend to gather. The crack where the ground meets the house is particularly important, since this is often used as both a hiding place and a nesting place. By treating this crack, preferably using the pinstream nozzle setting, you're applying insecticide where it is most needed, and where it tends to last the longest.

Keep your eyes open for any other cracks and crevices. Most homes also have cracks where the door threshold and sidewalk come together. There might also be cracks where the siding ends at the bottom of the house.

The width of the spray pattern depends on your particular circumstance. Contrary to what many people think, there is limited value to spraying wide swaths, and usually almost no value to regularly spraying the entire yard without a specific reason. Once you have the "average" home under control, a pint of liquid concentrate should last a year or more, even in the warmest climates where there are pest problems year round. (I live in Arizona, where

The crack where the ground meets the house is particularly vulnerable because many pests hide and nest here. Insecticide sprayed here will also last longer.

there is an abundance of very tough pest problems, and am just now getting to the bottom of a pint I bought four years ago.)

The need for liquid spray inside depends entirely on your particular circumstance. It's rarely needed except for pests that live inside the home or during the early stages of a regular regime of home pest control. Most pests can be stopped outside, before they enter the home.

Routine inside treatment should consist of nothing more than spot treatment, along doors and windows and at the baseboards where pests are regularly seen. There is no need to indiscriminately spray everything in sight on the chance that a pest *might* touch the chemical.

This is where a quality spray tank comes in handy. The fan spray makes it easier to apply a narrow strip of spray only near the baseboards without getting the chemical into the room where people—especially children—and pets can contact it.

In any case, use fairly low pressure in the spray tank, with the nozzle adjusted for a narrow cone. This gives you better control. And use as little spray as possible.

Back outside, use dust in dry areas, and on plants that have been bothered. Remember the basic rule—if you can easily see the dust, you've probably put on too much. The exceptions are when you put down a bead of dust in special places, such as in front of doorways and around ant hills.

If you have areas of moisture around the house, you may have to follow up the liquid spray with an application of granules. As always, use an amount that isn't obvious when you're done.

The job always ends with a thorough cleaning of the equipment, and a check to be sure that any and all chemicals are stored safely out of reach of pets, children or anyone who might accidentally poison themselves.

In most cases, the job should take no more than 30 minutes for an average-size house. Unless the pest problem is extensive, you probably won't have to repeat it more than once a month, and less often in the colder months.

You might find that one complete treatment a year is sufficient. Other treatments can be applied in specific areas periodically.

A Whirlybird makes spreading granules easy. For best results, walk at a steady pace while turning the crank at a constant speed.

5

Insects
General Information

As mentioned in the introduction, there are over 700,000 species of insects in our world, with some 10,000 new species being discovered every year. The variation in size, color, and shape is enough to boggle anyone's mind.

How can anyone keep them all straight? And, how are true insects separated from other tiny creatures? Ticks, mites, spiders, and scorpions aren't insects. Neither are centipedes, millipedes, or sowbugs.

Scientists use a system of classification. Every living and nonliving thing on earth fits into this system. No two names are the same. Once the name is given to a particular thing, there can be *no* doubt *exactly* what is being talked about.

The two major *kingdoms* of living things are plant and animal. The *phylum* gives a very general description of the animal. Humans belong to the phylum Chordata, along with every other animal with an internal backbone. Insects and spiders, and everything which has its skeleton on the outside are members of the phylum Arthropoda.

Classes narrow the possibilities further, followed by *family, genus, species, subspecies,* and so on. As the name grows longer, the description becomes more accurate, until it eventually defines one single kind of creature.

To see how it all fits together let's look at the full names of a few pests you might encounter:

Mosquito—Animalia Arthropoda Insecta Diptera Culicidae Culex pipiens

German Cockroach—Animalia Arthropoda Insecta Orthoptera Blattidae Blatella germanica

American Cockroach—Animalia Arthropoda Insecta Orthoptera Blattidae Periplaneta americana

Tick—Animalia Arthropoda Arachnida Acarina Ixodidae Rhipicephalus sanguineus

Black Widow—Animalia Arthropoda Arachnida Araneae Theridiidae Latrodectus mactans

You'll notice that the first three belong to the class of Insecta, while the tick and black widow are Arachnida. This tells you that, other than being animals with an exoskeleton (Arthropoda), the first three have nothing in common with the last two.

It also tells you a few similarities that they *do* have.

All members of Insecta have six legs and three main body segments as adults. The three parts are the head, the thorax, and the abdomen. In most cases, the head has two antennae, a pair of compound eyes, and sometimes auxiliary simple eyes.

The three pairs of legs are located on the thorax, as are the wings, if they are present at all. The abdomen is the breathing and reproductive center.

Metamorphoses, or growth changes, occur in one of two ways. The incomplete (or gradual) metamorphosis is similar to that of humans. Out of the egg comes a nymph, which closely resembles the adult except in size. In most cases, this nymph lives in the same place as does the adult, and eats the same food. It gradually becomes an adult.

A complete metamorphosis begins with the egg. The egg hatches into a larva (like caterpillars and maggots). After a period of time the larva enters a pupa stage, often weaving a hard case (cocoon) around itself. During this period of apparent dormancy the larva loses its worm-like appearance and becomes a full-grown adult.

Flies have this second type of metamorphosis. When you see a small fly, don't feel sorry for it because "it's a baby." What you see is the adult. It will never get any larger.

Arachnids differ in two respects from insects. Instead of having three body parts, they have only two. The head and thorax are a single fused piece. All the functions of an insect's head and thorax are carried on here. The abdomen performs the same function in arachnids as it does in insects.

The second difference is in the number of legs. Insects as adults have six; arachnids have eight. In some species, the young arachnid may have only six, but will gain the other two after a few stages (instars) of growth. Each time it outgrows its "shell," a molting takes place, like a snake shedding its skin.

Both insects and arachnids can be found worldwide, from the driest desert to the wettest jungle, from below sea level to the high mountains, and in an unbelievable range of temperatures. Their nests are found in the ground, in woods, trees, and plants, under or on water, inside animals, and a myriad of other places. One unusual variety of acari has been found thriving in a caustic solution of hydrochloric acid.

An insect's food is almost as varied; everything from microscopic plant and animal life, to a full-grown elephant. (The infamous army ant, although totally blind, can kill an adult elephant and strip it to the bone in three days.)

In size, insects can be anything from the microscopic to huge creatures like the atlas moth of India, with a wing span of one foot, or the walkingstick from the same area which attains lengths of 15 inches. Certain species of the lowly cockroach get to be more than 6 inches long. (Relax. They aren't in this country.)

Colors are usually "earthy," but can be startling reds and blues. There are stripes, spots, patterns, and designs that would make Rembrandt jealous.

The truth about insects can be stranger than your wildest dreams. Even the common insects you come into contact with are amazing. Despite all of man's effort to control or eliminate them, their numbers just haven't dropped. While we've managed to make hundreds of species of larger animals extinct, our efforts seem only to encourage the insect. If we give them adverse conditions, instead of leaving, they adapt themselves to suit it, often within just a few generations.

Control can be an exasperating business because of that peculiar ability to adapt. But keep at it, and follow your common sense, and you'll be able to at least control the majority of the pests around your home.

6

Control

Now we begin the actual operations of applying the insecticides to control your pests. In this chapter they are all listed in alphabetical order, instead of by classes and families. All pests you might come up against are here, including spiders and rodents.

HOW TO USE THIS CHAPTER

If you know exactly what your pest is, simply turn to that section. If you're not sure, each section has illustrations to help you make positive identification.

Carpenter Ant

Ants

Ants are perhaps one of the most interesting of all insects. They seem to hold an automatic fascination for children. And, they are one of the very few insects which people keep in the house as "pets."

A teaspoon of sugar and an ant hill can provide many hours of entertainment. For a while, the sugar might be ignored. Then one of the workers finds it with his sensitive antennae (or bumbles across it). Soon the treasure is known to the whole colony, and a steady stream of ants falls on the sugar, carrying it underground to be enjoyed by all.

Ants belong to what is considered to be the largest *order* of insects. Hymenoptera. (Others in this large group are the bees and the wasps.) Their metamorphosis is complete.

Ants are of the family Formicidae, which describes the formic acid they carry and inject their victims with.

Ants can be found almost everywhere in the world, from high mountains, to arid deserts, or steamy jungles. Size varies from as small as $3/100$ of an inch to more than $1\frac{1}{2}$ inches.

Although coloring is usually a shade of red, brown, black, or yellow, certain species sport bright metallic colors, like tiny, foil-wrapped Christmas presents. This coloring comes from two things; the integument, and (to a lesser extent) from the minute hairs that cover their bodies.

Life expectancy varies greatly between the different species. Some live for only one month. Others may live up to 15 years. The colony itself can survive much longer than that, as new queens replace the old.

Ants are social insects, with a highly-developed caste system. The newborn ants have little choice as to what they want out of life. Once they are born, they are stuck with whatever job nature equipped them for.

Most numerous are the worker ants. These are the ones most often seen as they comb the ground for food, or repair the nest. Workers are infertile females.

Some colonies have a special class, that of "soldiers." Armed with huge and powerful pincers, it is their job to defend the colony from enemies. And, just like man, the soldiers might also lay siege to other ant colonies, or termite colonies.

Fertile females are relatively rare. Most colonies allow only one queen. A few species have several, each coexisting on friendly terms with the others. The queen's sole duty is to lay eggs, which she does quite well. For her to bring forth 1,000 eggs per day is not at all unusual. (That's nearly $\frac{1}{2}$ million a year.)

After the fertile male finishes the "nuptial agreement," he is no longer required, and usually dies shortly thereafter. Then, his bride fills with eggs, often becoming so large that her feet can no longer touch the ground. From this point on she has to be waited on, mandible and foot. If the nest is disturbed, the workers gather around and roll her to safety. Quite unbefitting a queen, but necessary to the survival of the colony.

The average ant colony can survive the loss of nearly 80 percent of the workers and soldiers, but without their swollen, worm-like queen, the colony would probably perish.

Ants have several "human" characteristics. Not only do they go to war with their neighbors (usually over a boundary dispute), but they also can come back from the war with slaves.

In their invasion of the enemy's nest, eggs are carried away and brought home. When these hatch, the newborn become slaves. Quite often, if you see different species wandering unmolested around a nest, they are slaves.

Some ants exist only because of their slaves. These warrior ants develop such large mandibles that they cannot eat without help. Their slaves do all the menial work around the nest, and serve their masters willingly, not seeming to notice that anything is unusual.

Occasionally there may be a "slave uprising." If the master becomes too rough (which can easily happen, frequently by the master biting the head off the slave thinking its his food), the slaves rally together against their masters. If the slave-to-master ratio is large enough, the slaves might gang up and literally throw the master from the nest.

If the "master" has superior fighting capability, and even if they don't, these revolutions rarely last. Within an hour or so everything returns to normal, as if nothing ever happened.

Most ants have lost the power of their bite so that it is little more than a pinching. A few have not only retained that power but have modified their "equipment" for delivering the poison. One variety has the ability to "shoot" its poison at a target as far as two feet away.

One particularly virulent type is known as the fire ant. A decade ago fire ants were almost unknown in the United States. Now they are all too common throughout Mississippi and a few other southern states. Their bite is painful and can be dangerous, particularly because of the tendency of these ants to "gang up" on any intruder. They have been known to kill small animals, and to seriously injure larger animals—including people.

Treatment for the fire ant is the same as for other ants. The difference is that great caution is needed. Even walking too near one of their hills has been known to trigger a very fast and massive attack.

An ant's chief weapons are his pincers. In fact, these powerful mandibles are multiuse tools. The ant uses them much as we use our hands. They can also be employed for biting, piercing, tearing, cutting, to carry or build, and for a multitude of other things. About the only thing they aren't used for is eating.

The ant doesn't consume solid food, as such. Using his mandibles, he squeezes the "juice" from the morsel and drinks it.

This food can be as varied as the ants themselves. Anything humans eat, ants will also. Each species has a preference, but in a pinch (no pun intended) it will readily change its diet.

Most ants forage for food. Some are farmers, and raise different fungi for harvest. Still others are cattlemen, and keep large herds of "ant cows." They protect these, and "milk" them at regular intervals.

The "cattle" are usually one of several species of aphid. Sometimes scale insects are tended. The sap sucking aphid changes the plant juices to a digestible honeydew, which the ant relishes.

In certain species of ant there has developed an additional caste whose purpose in life is to store quantities of the honeydew for hard times. In parts of Arizona and Mexico, people search out the nests of these ants. The "caske" ants, being little more than bags of honey, are considered a delicacy, just as the honeycomb of a bee is.

The ant's sense organs are found mainly in its antennae. Where the termite has straight antennae, the ant's are elbowed, and sweep the air like radar scanners. Scientists believe ants sense things in geometric shapes. That is, to an ant, a certain distance might appear to him as a square, or a triangle. Scents might be recorded as circles, squares, and so on, and having qualities of hardness and softness.

The discovery of a food source is communicated to the other members of the colony. As much as it sounds like science-fiction, it apparently works very well for the ants.

No account of ants can be complete without a few words about the legendary army ants of South America and Africa. The army ant is totally blind. He relies on his sensitive antennae to tell him where he is going.

Army ants are nomadic, and have no actual nest. Instead, like a brigade of soldiers, they move to wherever they can find food, and rarely stop to rest. Like most meat-eating ants, they are primitive.

They aren't too fussy about what they eat. Anything that happens to be in their path is fair game. Injured or trapped animals are doomed. So voracious are the little creatures that they can kill an adult elephant within a few hours, and strip his bones clean in less than three days.

It's easy to see why the natives fear them. If a column is spotted, entire villages will be deserted until they've gone. If a goat or cow is forgotten, the villagers will return to find a pile of bones.

One good thing can be said for their invasion. Along with what damage they do, they also leave the village clean of all pests such as rodents and insects.

Generally, ants are considered more of a nuisance than a dangerous pest. Still, they can inflict painful bites, and might contaminate food to a degree.

The best control is to find the nest and poison it directly. As mentioned before, the colony can survive with the majority of the workers gone. Within a short time these will be replaced from the nurseries. By poisoning the nest itself, you stand a better chance of killing the queen.

If the nest is found, place a circle of dust around it, a few inches from the opening. The workers will then carry the dust into the nest on their feet and body hairs.

Granules also work quite well. Because of their nature, granules are activated by water. Once inside the moist nest, (the ants are more than willing to carry them down) the poison goes to work.

Liquids will also work. The ground around the nest should be thoroughly saturated with the chemical. You may have to retreat. And, don't be surprised if the only effect you see at first is the ants coming up in a new nest a few feet away.

Most insecticides will work on ants. Chlordane and heptachlor used to be the most commonly used and were highly effective.

Currently, diazinon is used most often, and can be purchased as a liquid, dust, or granules. A little harder to find, but also effective, is baygon, which is most often used as a bait.

To keep them out of the house, seal any holes or cracks. Also, foods should be kept wiped up from counters, and any spilled sugar must be cleaned up immediately.

Some ants make their nests inside walls or under the home. A small hole punched in the offending wall with a nail gives you access to inject the poison. (Be sure to seal the hole afterwards.) It is my experience that dusts are most effective here.

If the ants are found crawling along cracks on the counter, several different things can be done. The safest is simply to seal the crack, especially if the ants are nesting there. If you exercise great caution, small amounts of chemical can be applied along the crack. A cotton swab works well for this, and gives you better control of the chemical.

If the poison gets anywhere that it might possibly contaminate food or utensils, be certain to wash it off immediately. A few ants are easier to put up with than a case of poisoning. By using common sense and caution it might take a little longer to get control, but the additional safety is well worth it.

Green Apple Aphid

Aphids

There are many types of aphids, almost all of which do considerable damage to fruit and vegetable crops, and to flowers. They have piercing/sucking mouthparts, which they sink into the plant and drain it of its life juices. The leaves might then wither and die, and the buds, flowers, and fruits of the plant malform, if they form at all.

Other names for this pest are plant lice (for obvious reasons), and ant cows. (See "Ants.") Ants are particularly fond of them because aphids change the "cane" sugar of the plant into a more usable invert sugar. This substance, which is like honey and is sometimes called honeydew, is relished by many types of ant.

In a few cases, just getting rid of the ants will solve the aphid problem. The citrus aphid becomes so dependent on its ant protectors that it dies without them. They secrete so much honeydew that unless the ant is constantly tending them, they will literally drown in it.

Along with its attraction to ants, this honeydew promotes the growth of a black mold, similar in appearance to the mold that appears on old bread.

Certain species of aphid take on a dusty, powdery appearance because of the waxy honeydew. Others range in color from a lime green, to black, to a bright red. Still others are spotted or striped. Aphids may be winged or wingless.

There are quite a number of insecticides which help control aphids. Because of their rapid breeding and short incubation period, it often seems a hopeless battle. But keep at it.

If you pay attention, you will find that they are in the heaviest concentrations beneath the leaves near the veins, and close to the buds. Just as the tick seeks out blood vessels, the aphid has a talent for finding the best spots for its feeding.

Consequently, you'll probably have to lift the leaves to spray under them. A light mist is best when using liquid. Try not to drench the plant. Too much chemical could harm the plant more than the aphids.

In hot climates where the sun beats down like a nearby furnace, dusts are safer. Again, a light coat applied more often is better than a heavy one all at one time. A good rule of thumb is "if you can plainly see the dust, it is too heavy."

On vegetables, be sure that the insecticide isn't applied too close to harvest. The labels usually tell you how long before harvest you should stop applications. On any plant, avoid the use of aerosols. Aerosols are oil-based, and *oil* can kill the plant if applied too heavily.

Pyrethrum is a good "knock down" chemical, but has no residual. You can use it right up to the time you harvest. However, finding this insecticide in anything but oil-based aerosols is difficult.

Control

If this is the case, try using rotenone.

Diazinon and malathion are both excellent residuals, with Sevin running a close third. All three can be found in liquid and dust formulations. Granules are useless.

Still another reliable, and safe, chemical is methoxychlor. "Rose dust" usually contains methoxychlor as its active ingredient.

For those of you who prefer to garden "organically," there are three very good predators of aphids. The famed ladybug is an active little creature with but one fault. Apparently bored with her surroundings, she may decide to take an extended leave of absence.

The praying mantis love aphids, and are quite merciless. They are also entertaining to watch, and will be fearless of any ant guards that might be around.

Both the adult and the young of the pretty lacewing will eat aphids. If you can keep them around, they can be of great benefit to your garden.

Obviously, if you are using a natural predator to control your aphids, don't use insecticides. The poisons will kill both the aphid and the predator.

Bedbugs

Bedbugs have been plaguing mankind for millennia. It is believed that they made a pest of themselves with cavemen. The ancient Greeks and Romans were also bothered. Fortunately, with modern techniques of sanitation, bedbugs aren't quite the problem they used to be.

Bedbugs are Hemiptera, as are all true bugs. They have piercing mouthparts with which they drink the blood of their unwilling host.

The genus name, Cimex, was given to them by the Romans.

They believed that mixing a ground-up bedbug with certain liquids and drinking it would cure a snakebite and a number of other ills. This practice persists in some parts of the world today.

Bedbugs almost became of benefit to man, at least in a small way, during the Viet Nam era. It was found that some bedbugs emit a sound somewhat like a bark when they sense a human nearby, and get ready to feed. A group of scientists wanted to put this fact to use by encasing bedbugs in small capsules along with highly miniaturized radio transmitters. The idea was that the capsules could be dropped into enemy territory and monitored. If the barking was heard, the United States troops would know where to attack.

Their living quarters are by no means limited to beds. Any crack will do. They have been found in floors, walls, furniture, and in ceilings. They will move from house to house in belongings (including laundry), or may simply "walk over" to a nearby house for an extended visit.

Unlike many insects which drop their load of eggs at one sitting, the female bedbug deposits hers leisurely, a few at a time, until an average of 200 are laid. In about a week, the unfortunate host is then treated to a new population explosion.

The metamorphosis is incomplete, which means that the adult grows a little at a time, shedding its skin after a feed. After each molt the insect requires an increased amount of blood to satisfy its growing body. When fully grown, the adult bedbug is about $\frac{1}{5}$ inch in length.

Although they are only minor characters in the spread of disease, they are definitely unwelcome guests, especially to those people who are particularly sensitive to their bite. This bite can raise welts which itch like crazy. Scratching the wound only causes an open and infectious sore.

Treatment for bedbugs is not easy. Their nesting spots are so varied that finding them might be difficult. Spray all cracks from floor to ceiling. They could be hiding in any one of them. Use a residual, such as malathion or diazinon. If the problem is bad, you will probably have to use an aerosol flushing agent in the room.

Never spray the bedding or mattress. Remember that every night as you sleep on that bed you will be coming into contact with the poison. Keep the treated area to a minimum.

It's much easier to begin a policy of strict sanitation. Wash bedding and clothes thoroughly. Repair any cracks, holes, or loosened wallpaper. If you have a wood floor, it is possible that the bugs have made a home for themselves in the cracks. A new coat of varnish should help.

Carpenter Bee

Bees

The fear of "killer bees" has become almost a fad in our country. In truth, these bees are only slightly more dangerous than a honeybee. But the tales of their sting, and their highly protective nature, have given them (and all bees) quite a reputation.

If they *do* manage to reach the United States, they won't be able to survive in any of the colder areas. Even in the warmer areas where these bees can survive, most experts agree that there will be far less danger than reported by the "popular press." The bees normally do not attack unless they feel that their hive is threatened. Most of their aggressive behavior is directed toward other bees. The experts point out that even if the so-called African bees manage to drive out all of the present "domestic" bees, there is still no threat as the new bees will take over the important task of pollinating crops.

All in all, most of the worries about these bees are unfounded.

The vast majority of bees and wasps are beneficial to man. They usually sting only when threatened. Beekeepers will testify that it is extremely rare for one of their charges to sting them, even when they are close to the hive.

I have often stood within inches of a nest, and have been stung only once, and then only because I leaned back against the bee, crushing it. You can hardly blame the bee for striking back.

This does not mean that you should stick your nose into a bee's nest. Although the sting isn't dangerous for most people, it can be painful. Don't take chances.

An ancient fable tells how the bee was dissatisfied with her lot in life. She longed to be quick, like a cheetah, or strong, like a bull. Her constant complaints became such a nuisance to the gods that they decided to grant her wish, and gave her the power of a potent sting in her tail. Delighted, the bee flew off to try out her newly gained strength on the other animals. Seeing her intentions, the gods made a slight change. The sting would remain. But, they warned the bee, it could be used only once, and that once would take the life of the bee.

The barbed stinger, poison sac, and muscles which inject the poison, are all connected. Once the stinger is forced into the skin of the enemy, the entire apparatus is torn from the bee's body, and it dies within minutes.

The poison is a formic acid derivative, quite similar to that of its relative the ant. It acts as a neurotoxin, affecting the nervous system of the victim. (This poison has been used in the treatment of arthritis.) If the stinger is left in, the muscles which surround the sac drive the stinger deeper into the skin and drain the entire contents of the sac.

A clean fingernail or sterile knife blade can be used to scrape away the sac and muscles before this can happen. If nothing clean is handy, leave it alone, unless you relish the prospect of infection on top of the sting. *Never* slap or rub the wound or attempt to pull the stinger out. This will only squeeze the poison into the wound.

A poultice of bicarbonate of soda can be applied. Another remedy suggests using finely chopped onions. This not only draws out the poison (by osmosis), but can also create sufficient "suction" to remove the entire stinger. (I've used it myself, and it works!)

Despite what you might think, most bees are solitary creatures. Very few are social insects. The carpenter bee, which can be responsible for considerable damage to wood because of its habit of burrowing into wood to make its nest, is found both singly and in small groups. These bees are large and dark, usually with a shiny, almost metallic coat.

Honeybees and bumblebees are the most commonly occurring bee pests. Both are social, and build colonies. The queen may lay up to 1,800 eggs per day.

Women's Lib was practiced by bees long before people ever dreamed about it. The male serves only to fertilize the queen. After this, they are driven from the hive or are killed. The workers are all infertile females. They also serve as protectors of the nest (unlike termites, which usually have a special soldier caste).

Caution should be observed when attempting to control any kind of bee or wasp. First you must find the nest. If there is none, or seems to be none, you'll just have to resign yourself to their presence.

Dusts are usually best, with liquid being used more for a quick contact kill. Bees are most active during the day, so try to dust the nest at night. Pump the dust into the opening, and it should completely eliminate the problem.

A little safer, but less effective, than using a duster is to use the pin stream of your pressurized tank. This way you can stay at a distance from the nest.

If you don't have protective equipment, either leave the job to someone who does, or proceed only if the problem has become intolerable. And then, use every precaution. One sting isn't too hard on your system, but a swarm of stings can prove dangerous, even fatal. To someone allergic, even one sting can be serious.

If the hive is in the open, such as in a tree or bush, try to contact a local beekeeper who might be willing to come and take it away for you.

Beetles

There are more species of beetle in the world than there are of any other insect. Many are scavengers, feeding on anything dead, including leather, stuffed animals, and stored grains. Some prefer live plants, and cause millions of dollars of crop damage every year. Despite the damage a few species bring, the great majority of beetles are beneficial.

There are very few places on earth where beetles won't live.

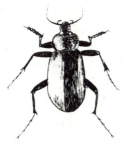

Ground Beetle

There are tree beetles, ground beetles, and beetles that live underwater. Even if the outside climate is too harsh, they can be found inside the home, in boxes of cereal, or in other stored products.

Beetles range in size from the almost microscopic to huge beetles such as the goliath beetle of South America which attains lengths of 5 inches and weighs ¼ pound.

While many beetles are black and brown, others can be beautiful contrasts of bright colors in stripes, spots, and designs only the hand of God could paint. Occasionally, you might come across one that seems to be clothed in shiny foil, as though a child were trying to play a trick.

One beetle, the bombardier beetle, is rare in that it is one of the few things that can survive among a host of army ants. It wanders along with the troops, sometimes snatching a dead insect from the ants' hungry jaws. If the ant comes too close, the beetle raises its tail and releases a cloud of foul gas. Back in the ranks (after beating a hasty retreat) the unfortunate ant is regarded suspiciously for several days until the scent wears off.

The rhinoceros beetle looks like the insect world's version of a triceratops. This incredibly large beetle has a huge sprout on his head, like the thorn of a rose.

All beetles have wings, although many seem to never use them. The tiny black darkling beetle, common in Arizona, proves to be a plague to exterminators because they cannot convince their customers that the persistent beetle flies. As with all his relatives, the flying wings are hidden beneath a hard shell (which is also a pair of wings, used for gliding).

Because of their ability to fly, and their fondness for hiding, beetles are often difficult to control. In many cases, a mere decrease in their numbers is the best that can be hoped for.

Dusts or liquids can be used. Granules are usually a waste of time, as are baits. Diazinon or carbaryl are probably the best poisons to use. At times, silica also proves to be very effective.

In the garden, spray or dust the plants and ground, hitting directly as many beetles as possible. Repeat treatment as often as necessary, depending on the extent of the infestation.

To keep them out of the house it will be necessary to plug any holes or cracks. Although beetles fly, and are often attracted to light, the most common way of entry of outside nesting beetles is to crawl under badly sealed doors.

Mobile homes present a special problem. Because of their construction there are literally thousands of possible points of entry. By dusting under the home, and then spraying the inside with a residual, you should attain a fairly satisfactory result. Pay extra attention to doorways, windows, and vents.

As pantry pests, snout-nosed beetles are without equal. (*See also* "Weevils.") When beetles are found in flour, or any stored product, the first thing to do is throw out all of the contaminated material. It's useless to try to sift out the insects. Even though you can get out all of the adults, their eggs are too small and will remain to reinfest.

After you've done this, empty and wash out the cabinets. Then, treat them (carefully!-only along the cracks) with residual. Be sure to get the chemical into the cracks. This is where the pests will be hiding.

Finally, "quarantine" any suspicious products, as well as any incoming food. Tupperware, or a good jar will work fine. The idea is to keep the pests from spreading into other foods. One box with just a few weevils can infect everything in the cupboard in a remarkably short period.

Black Widow

I still remember my first encounter with a black widow. We were visiting family friends, who live just outside Jackson, Mississippi. Being raised in icy Minnesota, I had never even heard of black widows. One morning, as I was exploring the grounds around the

house, I came across a dead lizard. The only wound he had was an odd-looking blister at the base of the tail. In his mouth was a small bit of black flesh.

Curiosity set in, and I went looking for whatever could have made the wound on the lizard. A few feet away I found it. Apparently the hungry lizard picked a black widow for his breakfast, and she had justifiably protested. Her abdomen was torn open. So, although the widow hadn't escaped her fate completely, the lizard was denied his last meal.

I still have that spider sealed in a small jar of alcohol as a souvenir of my visit to the South. I never expected to see another black widow. But fate stepped in, and I ended up living in Arizona, and have since handled more than my share of the little bundles of black death.

When people find them around their houses, especially if they have children, they don't mind paying the exterminator. Although the bite is rarely fatal to adults, there have been a number of deaths in children attributable to it.

The poison is an extremely concentrated neurotoxin. It goes to work almost immediately, with a spreading dull pain. Within minutes the victim's body protests by profuse sweating, a feeling of nausea, and sometimes convulsions, unconsciousness, or (rarely) death.

If you ever should get bitten, *don't panic*. Immerse the wound in ice or cold water. This will slow the poison's entrance into your body and will also help cut the pain. *Never* cut the wound open. There isn't enough poison to suck it out. (Even in the case of snakebite, this famous technique is foolish.) If at all possible, get to a doctor immediately. A black widow bite isn't usually all that dangerous, but it certainly isn't worth risking your life.

Despite its reputation, the widow is a mild creature, and usually requires considerable provocation to bite. Trouble is, you can't ever be sure what might provoke one. Many times they'll crawl along your arm for hours, perfectly happy. You can pet them, push them along with your finger, and even press down on them with no results. Other times, a finger just coming within range of their powerful fangs is sufficient.

The adult female is jet black, with a somewhat spherically shaped body. When high buttoned shoes were in fashion, they were given the name of "Shoe Button Spider" because of their shape.

The legs are large and powerful, with visible joints. Altogether, the spider can grow to widths of 1½ inches. (The abdomen rarely exceeds ½ inch, and then only when gravid.) Under the abdomen are the two red triangles, joined point to point, making an hourglass shape. As the spider normally hangs upside down on her web, the design is seen only when looking at her from above, or when she turns.

A few words of caution in identifying the widow. She is usually black, but not always. The red marking on her abdomen is usually a red hourglass, but not always. It may be merely a splotch, or completely missing. Coloring can range from grey (a subspecies) to a blackish-red, to even a grey-yellow. These variations are rare, however, and are not passed on to the spiders offspring.

In many cases, before you see the spider, you'll find what appears to be an empty web. Her web is unique, and entirely amazing. Each strand has 10 times the tensile strength of steel! When you move a stick (*don't* use your hand) through it, it crackles audibly as the strands break. But, the remarkability doesn't stop there.

Once more she outdoes man by making each strand almost perfectly uniform. So perfect are the strands, that man "farms" them for use in fine optical equipment. During World War II, farming black widows for their webs was a big business. The webbing was used for the crosshairs of bomb sights.

The web has no pattern. It is completely erratic, running every which way, and connecting wherever the spider happened to be at the moment. But, it serves its purpose. So sticky and strong is it, that full-grown mice have been trapped in it.

As all spiders, the widow is a predator, feeding on insects that stumble into her web. Her eyesight isn't particularly good, but the vibrations in the web alert her to the intrusion. She then spins a strand and approaches the victim backwards. If the insect is large, or moving too much for her liking, she throws drops of unspun web on the insect. These drops glue the insect in place, and its thrashing only entangles it more.

After the meal is finished, it is usually cut loose and allowed to fall from the web. The widow is an excellent housekeeper. Generally, if the web is dirty in appearance, with debris hanging from it, then there is no spider in the web.

Eggs are laid one at a time, then sealed in a cottony sac which looks very much like a white berry. These eggs (up to 1,000) hatch within three weeks. The spiderlings begin feeding on each other, thinning out the "weaklings." In about a month, the sac breaks open and releases its contents of white spiderlings, each about the size of a pinhead.

As they grow and molt their outer skins, color begins to develop, first as dark stripes, then slowly spreading into the full black of the adult. At certain stages these markings can become quite colorful, with greens and reds and yellows all intermixed. Occasionally the marks don't fade, and remain even at maturity.

The spread of black widows occurs in two main ways. If the original web is in a closed space, the young move away from the main web and build their own extensions, each generation moving a little further than the previous, much the same way flowers do.

The second way is with the help of the wind. Being almost weightless, the spiderlings are easily picked up by a breeze and carried away. They drop a length of webbing, which acts like the tail of a kite, and serves to catch onto whatever object they come across.

Homes along the path of the breeze find themselves deluged with the spiders. Sometimes the area between two houses will be covered with black widows while the far sides will be clear.

When given a choice they prefer to make their webs where it is dark, damp, and secluded. They are shy, and would rather not be disturbed. Also, food is found in greater abundance in these areas. Usually the webs are against objects, like the sides of buildings or boxes. At times they will make their home under a rock or board, in which cases there are no webs.

As always, control begins by cleaning up the yard. Every bit of debris, every box, or bush, or woodpile gives them a place for a home. Objects that jut up provide a nice spot for the webbings of wind-blown spiderlings to catch onto, as will any old webs. As you clean the surroundings, break down any webs.

Diazinon is excellent. Dust attics and crawl spaces. Fences, bushes, and other objects should be dusted or treated with liquid residual. Also spray along the perimeter, and under awnings, and eaves. If your home has a shake or tile roof, or loose shingles, spray or dust there also.

Inside, pay special attention to basements or any dark, undisturbed places. Spray behind appliances, especially in the laundry room. Closets and storerooms (inside and out) are popular nesting sites.

Pyrethrum, in an aerosol, can be used for a quick kill to flush the spiders into the open. In badly infested spots it might be necessary to fog the entire room. It'll chase them out of their hiding spots, so keep your eyes open.

As mentioned before, break down all webs. If there are any egg sacs, destroy them. If the web is rebuilt, you'll know that the spider escaped your efforts. Hit the area again, and look for any holes or cracks you missed the first time. If the web is there, so is the spider.

If you live in the "back woods," or are on a camping trip where there is an outhouse, be careful. A common place for their nest is in the opening where you'll be sitting. The genitalia are an unfortunate and extremely dangerous place to be bitten. It's a good idea to brush a stick inside the opening before sitting down. This will knock down the web if it's there, and your rump won't jiggle the web, or entice the spider.

One final note. The male black widow is a rather insignificant character. He is much smaller than the female, and not at all dangerous. He lives only to fertilize the female. Even if he escapes her jaws after mating, he won't live much longer. The male is a light brown, usually with light streaks along the abdomen. He is *not* a recluse spider as some people think.

Box Elder

The half-inch red and black box elder is a garden pest. Although they make quite a nuisance of themselves by coming into the house, they don't eat anything in the home and they don't bite.

Actually, they do very little damage to plants. Their favorite feeding spot is the tree from which they get their name, the box elder. They will also frequent maple, ash, and apple trees. If none of these are present, they might feed on a number of other trees and plants. With their apparently small appetites, however, you'll hardly notice the difference.

The adult lays her eggs in the bark of the tree, or in any crack or hole. I've seen them depositing the eggs in the holes found in block walls. Their young hatch in the spring and begin feeding almost immediately.

Since they are good flyers, control is extremely difficult. It usually requires spraying large areas, and trying to hit directly as many insects as possible. Treating areas where they walk will also help a little, but not much.

Even with all this, it's almost a certainty that they'll be back. If you manage to kill *every* box elder around your house, within a few days others will fly in.

About the best chemical for them is diazinon. Spray or dust the trees and plants where they are feeding as well as any other places you see them. To keep them from coming into the house, be sure that the doors and windows are well sealed. A bead of dust along the outside of the house is good.

Good luck. You'll probably find that you'll just have to learn to live with them.

Brown Recluse Spider

The brown recluse spider is often mistaken for the male black widow. Actually, the two have nothing in common. The male black widow is smaller than the recluse and lacks the violin-shaped marking.

When fully grown, the recluse is about 1½ inches long, the same size as the female black widow. The legs are heavy and strong. Coloring is tan to brown, with a distinct "violin" on the top of the cephalothorax (the part where the head and legs are).

The webs can be found in the same sort of place as the widow's—dark, damp, and secluded places. Usually the webs are against objects, like the sides of buildings or boxes. The recluse, however, is even less bold, and is also quite rare. When they do come, it is sometimes in droves. Hundreds of the spiders can make their presence known all at once.

The bite begins with a slight pain. Within 12 hours the pain increases, and the wound infects. The complications of their bite can be serious, and may take several months to clear up. Even then it can leave an ugly scar.

As with the black widow, don't take any chances. Put ice or cold water on the wound. Then, get to a doctor. The bite of a recluse is no joke. If anything, it may be considered *more* dangerous than a black widow, because of the secondary infection that is almost guaranteed to set in.

Control is the same as with the widow. Treat any dark, secluded spots with diazinon in liquid or dust form. Unless the spot is overly damp, dusts will probably work better. Outside, remove any debris, such as boards, from the vicinity. These spiders get their name from their habit of living as "recluses." Only rarely do they come into the open.

Centipedes

Centipedes are not insects. They belong to the special and small class of Chilopoda. (Insects are of the class Insecta.)

Their bodies are segmented, with each segment having one pair of legs. The common house centipede has 15 pairs of legs. Other species have different numbers of legs, but never 100.

All centipedes possess poison glands, which are located just below the jaws. In most species, the jaws are simply not powerful enough to penetrate the skin of a human. Even if they did, the bite is weak.

There are exceptions. The large, reddish centipedes found in Western states can deliver a very painful bite. Fortunately, these are rare. And, like their smaller brothers, they are shy and will go out of their way to keep hidden.

The house centipede makes its home in damp places where it can find a supply of insects and spiders for food. When they aren't actively seeking dinner, they crawl under any nearby board or box. In the warmer months they seem to prefer the outside.

The desert centipedes nest in a variety of places. A favorite is under leaves or rocks. Occasionally, they may be found inside hollow walls, especially if there is moisture gathering in them.

To control centipedes you must find and destroy the nests. Make conditions bad enough and they'll move on to friendlier territory.

If the centipedes are in the wall, use pyrethrum or DDVP as a flushing agent. Aerosols, liquids, or dusts will work. Which you use depends on the conditions. (A damp wall rules out dust, for example.)

Liquid residual should be applied along the baseboards throughout the house, with extra attention to basements, closets, and dark cabinets. If there are any holes to the outside, patch them. Centipedes are tough to kill. Chemicals take a long time to work. And, if the centipedes have easy access to the inside, your spraying will be much less effective.

Spray or dust crawl spaces and attics. Around the outside remove any debris. Decaying vegetation provides an excellent nest and must be cleaned up. Be careful when you do this as there might be a centipede already there.

Diazinon or malathion may be used. If the ground is damp, you might want to use granules. Dust works best on dry ground or in wall voids. Liquids can be used for spraying the lawn and under bushes.

COCKROACHES

Now we come to what is one of the most prolific of all pests—the cockroach. There are many different species of cockroach. This chapter is divided into sections, each covering a specific type of roach.

Their order is Orthoptera, which closely relates them to grasshoppers and crickets.

Cockroaches are probably one of the oldest living things on earth. Through the centuries, they have survived by adapting themselves to whatever conditions faced them. Even now, they continue to change and adapt. The most obvious change is their relative immunity to certain pesticides.

The name cockroach comes from the Spanish word "Cucaracha," which more or less means "the crazy bug." The technical family name of Blattidae is derived from the Roman "blatta," or "flees from light."

Believe it or not, roaches *do* serve a useful function in nature. First, they are wonderful predators of bed bugs, and of several other insects. Second, they are basically omnivorous. They'll eat just about anything they come across. Consequently, roaches make themselves useful as scavengers, and help in the decomposition of many types of natural debris.

Certainly you wouldn't want to invite them into your home for these reasons, but in nature they perform a very valuable service.

Despite their lowly stature, roaches have been active participants in studies of space. Unlike monkeys, roaches need next to nothing in the way of life support. A typical cockroach space capsule is like a beer can with the top cut away. The six-legged astronauts are strapped in with a piece of cotton protecting them from the forces of acceleration. So, when you hear the news that a human has landed on Mars, a cockroach may have been sent first, to see if it's safe.

Very, very few species of cockroach come into the home for food or nesting. The vast majority live outside where they don't bother anyone but other insects. Some live in trees, others in or around water, and still others invade the nests of ants and other insects.

Whereas the house-invading types are usually brown or black, and small, the tropical roaches have an on-going competition with their cousins, the beetles. Coloring might be a bright, luminous green. One species of roach grows to a length of six inches.

As mentioned before, roaches are omnivorous. Most prefer sweets and starches, but they'll eat anything. They don't seem to care particularly. They'll devour plants (living and dead), upholstery, books, dead animals, cheese, beer, leather, and wood. (Some scientists believe that a cockroach evolved into a new line of work and became a termite.) One of the common roaches of South America has a particular fondness for human hair and fingernails. Entire villages may be found where the people have very little facial hair left.

The major objection to cockroaches is their ability to transmit disease. Among the diseases traced to cockroaches are cholera, salmonella, plague, and leprosy.

All cockroaches have an incomplete metamorphosis. The newly hatched young usually look similar to the adult (except in size), and go through phases of growth called instars. (If you see what appears to be an albino roach, it is merely a roach ready to shed its outgrown skin.)

American Roach

As with all the other "ethnical" roaches, no country wants to admit itself to be the origin of this red monster. Almost all come from the tropics and were brought to the "new world" aboard ships carrying cargo.

American Roach

We just happened to be lucky enough to have this one named after us. He is known by other names, however, such as waterbug, palmetto bug, shad bug, and by the very colorful title of Bombay canary.

The American cockroach reaches lengths of up to about 2 inches. They need a high level of humidity to survive. Because of the large body surface, body moisture evaporates quickly. One of their favorite habitats is in sewers or septic tanks. Water traps in pipes are no barrier.

Generally, they make other homes wherever they can find moisture. Around water pipes, in walls and ceilings, in garbage dumps, basements, and in the crawl spaces beneath houses, are all common nesting sites.

They spread by being carried in boxes or bags, by the sewers and coming up through the drains, and by flying. Within a building, such as an apartment building or office, they will merely walk from one place to another.

They live up to 2½ years. During this time they plod along, often making more noise than a wall full of mice, dropping capsules of 16 eggs each.

To be perfectly fair, the female American roach doesn't just drop the egg capsule. Actually, she takes a great amount of care, making certain that it is well hidden.

Since their nesting areas are rather limited, control usually isn't too difficult. Liquids, dusts, granules, and baits will work quite well. Treat all possible nesting spots. Spray or dust attics, false ceilings, and crawl spaces. Sometimes you may have to treat inside the wall. Diazinon, malathion, baygon, or carbaryl are all good chemicals for American roaches.

Baits should be placed near the nests where the roaches are sure to find them. There are several good commercial baits. One of the best contains baygon. But, one of the simplest baits is a partially filled bottle of beer. If you use baits, put them out of sight, and out of reach of children and pets.

If the roaches are coming through the drains, treatment is a little different. Most places have local ordinances against pouring pesticides into drains. The only time you can legally treat the drains is if you have a private septic system. But then, a strong pesticide might deactivate the tank. Be careful.

For city dwellers who are tied into a sewer system, the solution is simple. Contact the Department of Sewers and Sanitation. They will gas the sewer beneath your home without charge.

Australian Roach

The Australian roach is very similar in appearance to the American. It is somewhat smaller (about 1 inch) and also bears distinctive yellow markings on the thorax and sides.

It is an excellent flyer, and often makes its nest in trees (with a special liking for palms). They definitely prefer to be vegetarians, and turn to eating books and other plant products at times, which makes them a nuisance.

Where the American roach has only 16 eggs per capsule, the Australian has 26. But, their life expectancy is much shorter, being only about one year.

Occasionally, they may wander into the home. To help slow them down, spray along the baseboards, doorways, and windows. Outside, spray or dust the entire perimeter. A bead of dust along doors and windows may help too. If the ground is damp, granules will be effective.

To get a more positive control, you'll have to locate the nests. Dust any holes in trees as these are a favorite nesting spot. Check around the house for any other nests. At times, the Australian roach will make a home in an attic, crawl space, or garage.

Diazinon, Sevin, or malathion may be used, with pyrethrum being, as always, excellent for flushing and for a quick kill.

Brown Banded Roach

Unlike so many other species which have predictable nesting spots, the brown banded roach can be found anywhere in the home. This habit makes it one of the more difficult to control.

The adult is about half an inch in length, and similar in appearance to the German cockroach. The easiest way of telling the two apart is by the yellow, or light brown stripes on the brown banded. Also, the German roach bears two dark markings on the thorax, which the brown banded lacks.

Treatment for them can become a long, frustrating battle. Although they breed slower than the German roach, their fondness for picking unlikely spots for nesting makes eradication difficult.

Using diazinon, concentrate your attention on drier places. They are rarely found nesting near water. Other likely spots are in furniture, behind clocks and pictures, on and behind shelves, and so on. Brown banded roaches show a preference for high places, so your search must take in the entire room, from floor to ceiling. Dust or spray the attic.

If possible, use a fog to chase them from their nests. If they are spread throughout the house, this might be one time you'll have to call in a professional. He has the equipment to fog the entire house. If you decide to try it yourself, clear all animals (including fish) from the house. Once you've filled the house with the insecticide fog, get out and stay out for as many hours as possible.

Even then you may have to retreat. (No pun intended.)

German Roach

The most common roach in America is the German cockroach, sometimes called the Croton bug. This second name comes from the belief that it was first distributed through the pipes of the Croton Reservoir in New York.

The adult is brown, often with a greasy, polished shine on the back, and two dark spots just behind the head. For being only half an inch long, this roach is one of the most prolific of all roaches.

Each egg capsule contains an average of 35 eggs. In her life, the female may be responsible for the addition of up to 400,000 new roaches. The young reach maturity in as little as one month, and are then ready to begin their own campaign of procreation.

It's easy to see how these roaches have become one of the major infesting insects in the country. They are the "bread and butter" of exterminators. Normally, control is attained only after months, or years, of constant vigilance.

They spread in many different ways. They will ride in on boxes and bags, or will migrate from house to house by walking or flying. Once inside, they immediately set about finding a comfortable and safe home.

Kitchens are by far the most popular place. They crawl into every available crack and crevice. Heaviest concentrations are found near a source of water. Normal condensation on water pipes provides sufficient moisture for quite a crowd of German roaches.

When the population increases, the roaches will make their nests in the open, along the interiors of cabinets. (Normally, they avoid light and open spaces.) Occasionally, they may decide that an appliance makes a good condominium.

In one restaurant, I was ready to give up trying to control these roaches. I'd gotten rid of most of them, but they still showed up. I had checked and treated every conceivable nesting site. After a long and frustrating head scratch, I finally found the nest . . . inside a telephone.

They will also move into wall voids, behind sinks, and under any object (like a box, or pile of clothes) on the counter or floor.

Like most roaches they prefer starchy foods, with a strong liking for grease. In over-populated kitchens, they will turn to eating paper, and even the wood of the cabinets. More rarely, they may decide that human hair makes a nice change of pace for their diet. An open beer is like a dinner invitation for them. They love any fermented substances.

The first step in control is a thorough house cleaning. If you have an abundance of German roaches, you're doing something wrong. Lack of sanitation only encourages them to multiply. Floors and walls should be scrubbed to remove any grease. Don't forget to clean behind and under appliances. Not only does dropped food have a tendency to collect here, but these areas also provide seclusion and warmth.

Any holes or cracks must be sealed, especially when they occur around water pipes and lead into the wall void. Wash all cupboards, and replace the shelf paper.

At least until the problem is under control, don't leave any rags lying or hanging around. After the rag is used, remove it from the area and wash it. After it has dried completely, it can be brought back.

German roaches have developed an immunity to several insecticides. Chlordane has only a slight effect on them. Diazinon, malathion, baygon, and dursban are all good, effective residuals. For complete control you'll probably have to use a flushing chemical, either pyrethrum or DDVP.

Remove all dishes, food, and utensils from the cabinets so as not to contaminate them. These should be left out until the liquid residual has dried.

Using dust or aerosol, treat wall voids. (Liquids won't disperse properly here. Use the liquid to spray cracks, baseboards, water pipes, and wherever else you find the roaches moving.)

Baits may be used for a backup. By themselves they'll handle only a very small infestation. When placing baits, put them well out of reach, and sight, of children or pets.

Oriental Roach

These ugly, foul smelling, Oriental black roaches are considered the filthiest of all roaches. They grow to lengths of 1½ inches. Unlike the other roaches, they move slowly, and aren't a bit shy.

The Oriental roach finds its home in basements, beneath houses, in pipes and sewers, and any place else that is dark and damp. At times they will invade the kitchen and other parts of the house, and make nests inside appliances, especially ones which create moisture.

They are found in the greatest numbers on ground floors, but will climb. In time, and given a large enough population, they will make their way to the upper floors. They spread between homes primarily through sewer lines. Because of this, if one house in a neighborhood has them, it's a good bet that the others do, or will have shortly.

They feed almost exclusively on filth. Garbage and decaying matter are their favorites. Consequently, they are known carriers of disease.

Life span is just over a year, with most of that time spent in "growing up." After the roach reaches adulthood, it will live only a few months longer. Like the American roach, the Oriental roach's capsule contains 16 eggs. This, coupled with their short life span as egg-laying adults, makes their population increase slowly (fortunately).

79

Because they like to live in sewers, control is difficult. You may have to call in the Sanitation Department to treat the sewer lines.

If any water leaks are present, fix them. Weeds and decaying vegetation should be cleared away, especially around ventilators which come into the house. The drier and brighter you can make the home, the better. If you have a basement, keep it clean and dry. The same applies to crawl spaces.

Another help is to install tight-fitting drain covers with fine holes (the smaller the better). This will keep them from crawling into the home from the sewer.

Treat the inside with diazinon, baygon, or dursban. Spray along walls and baseboards, and inside lower cupboards.

Outside, spray or dust the perimeter. (Use granules if the ground is damp.) Along with the insecticides previously listed you can also use carbaryl outside.

Pennsylvania Wood Roach

The Pennsylvania wood roach is common in the eastern, southern, and mid-western states. They are 1 inch in length (the female is slightly shorter), and a chestnut brown in color with a tinge of white lining the thorax and wings.
is slightly shorter), and a chestnut brown in color with a tinge of white lining the thorax and wings.

There is considerable difference in the shape of the female and the male wood roach. At one time they were thought to be two separate species. She is usually one-quarter inch shorter than the male, and has only small, vestigial wings. The male has full wings, which he uses skillfully over short distances.

They are an outside roach, living in woodpiles, hollow trees, under loose bark, and in the cracks of homes. At times they will take over an abandoned beehive.

During the cold winter months the nymphs hibernate, but become active if disturbed. The cold has little effect on them.

The male has one characteristic unique to cockroaches. Instead of being repelled by light, he is attracted to it. Drivers are sometimes plagued by the male roach along rural roads at night.

Dusts seem to offer the best control, and should be applied around the perimeter of the house, in woodpiles, and on nearby trees. Diazinon, or carbaryl may be used.

Smokey Brown Roach

Smokey brown roaches are closely related to the American cockroach. (The American roach is Periplaneta americana. The smokey brown is Periplaneta fuliginosa.) They resemble their relative in shape, but are smaller (just over one inch) and are much darker, almost black.

The wings of both sexes reach out beyond the body. They are good flyers, and have been known to travel long distances in the air.

Each of the 17 capsules the female lays in her life contains an average of 24 eggs. In about two months the eggs hatch, and grow to full adulthood in a year.

Their favorite foods are fruits and sugars, but they will feed on a variety of materials. Their natural habitat is in woodpiles and trees. In the home they infest crawl spaces, attics, fireplaces, and roofs.

Since they fly, entry into the home can be through any opening large enough for them to squeeze through. They may also be carried in on boxes or wood.

Treatment is similar to that for the Pennsylvania wood roach. Seal all openings. Spray or dust the outside with diazinon, or carbaryl (dust is better). Also treat any nearby woodpiles and the cracks in trees.

Attics, crawl spaces, and gutters at the edge of the roof will have to be sprayed or dusted regularly.

Vaga Roach

The dusty, golden-colored vaga roach causes nearly as much frustration in the western states as the smokey brown and Pennsylvania wood roaches do in the east. Like their eastern relatives, the vaga roach is an outside insect, but will wander into the house. They are also excellent flyers, which makes them all the more difficult to control.

The vaga roach (also known as the field, or agricultural roach) is very closely related to the German roach. Its appearance is strikingly similar, even to the two dark spots on the thorax. The main difference is the coloring. Where the German is a shiny brown, the vaga roach is a dusty gold, and has almost white legs.

Although they will eat small amounts of food inside, they won't nest there. The German roach is found mainly in the kitchen area. The vaga roach occurs throughout the house, especially in rooms that have badly sealed doors or windows.

Their nests are outside in irrigated fields, citrus groves, and around the house where there is plenty of decaying vegetation and moisture. They feed on this vegetation, and on other insects.

The easiest way to discourage them is to clean away as much vegetation as possible. If there is a leaky faucet, the moisture helps promote a population increase. Remove their food and nesting spots and you won't be overrun with these roaches.

For chemical control, use diazinon, malathion, or carbaryl. Granules or liquid should be applied to the lawn. Dust or spray crawl spaces.

With a pressurized hand tank or garden hose attachment, soak the ground around the house. A favorite nesting spot is in the crack where the ground meets the house. Also spray or granulate under boards and rocks.

Inside, spray along baseboards and doors. The floor vents in mobile homes should be treated as well as window sills (inside and out). The better the seal on doors and windows, the fewer roaches will find their way inside.

If they *do* get in, despite all your efforts, don't worry. They do no damage to the home.

Crickets

In trying to imitate the "cri-cri" sound of this insect, the French gave it its common name of cricket. It belongs to the order of Orthoptera, and is related to cockroaches, grasshoppers, and the mantids.

Like grasshoppers, crickets have chewing mouthparts, and powerful back legs for jumping. Most crickets have wings, but these are rarely used.

Their main food source is plant life, or anything made from plants. They become a double pest by doing damage both in the garden and inside the house, where they may feed on clothing and carpets. They do eat other insects, however, and so aren't completely without virtue.

The cricket's song is produced by scraping the upper pair of wings together. As pleasant as this sound is to some people, it can become quite irritating on a sleepless night.

The Chinese have kept crickets in small bamboo cages for their chirping. In China, it is considered bad luck to kill a cricket.

Crickets serve another useful purpose in the Orient. Like dogs and roosters, crickets have been raised to fight in the ring. Two male crickets are placed together where they battle to the death. Sometimes large amounts of money are bet on the outcome.

Like cockroaches, they have an incomplete metamorphosis. Up to 400 eggs are laid, one at a time, in cracks, or in the ground. They prefer the outside, but will come in, especially when driven by cold weather or excessive moisture.

The house cricket is light brown with three dark spots on its head. The antennae are long and thin, and curve back around the body. While only ¾ of an inch long, it is a feisty little creature and will nip.

The field cricket is slightly larger and darker. Many different subspecies of this cricket can be found all over the world. They can cause considerable crop damage by their voracious appetites.

One of the more unusual varieties is the mole cricket. Its front legs are like clawed hands. With these well-suited tools, it burrows into the ground where it often uproots small plants.

Crickets are generally nocturnal. During the day they seek dark, damp spots, such as in bushes or in cracks. Homes near streams, ponds, or dumps may find themselves swarmed with crickets.

Tall grass, weeds, leaf piles, rubbish, under cans and decorative bark are favorite spots around the home. They are excellent climbers and might be found nesting along the roof and in ceiling beams.

The first step in effective treatment, as always, is to clean up as many potential nesting areas as possible. Grass should be clipped short. Clear away weeds and any decaying vegetation. Keep garbage in cans, and keep the cans up off the ground. Cracks and holes must be sealed up at all heights.

Baygon has almost miraculous results on crickets. Diazinon, malathion, and carbaryl are also effective. Which form you use depends on the circumstances. Wherever possible use dust.

Spray the entire perimeter. Chemical barriers around possible nesting sites will help to discourage them. Spray or dust all cracks and holes, as well as woodpiles. As with vaga roaches and earwigs, saturate the crack where the ground meets the house. Spray or dust the attic and crawl space.

Inside, spray along baseboards, doors, and windows. Basements, closets, and other dark places may require special attention. Also spray in vents and under furniture.

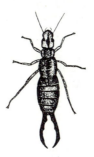

Earwigs

Almost everyone has heard the myth that surrounds the pronged creature the earwig. The tale goes (in any of several versions), that once upon a time a man awoke to find that some horrible creature had entered his brain through his ear. The horror in his head then began to lay its eggs, and when the young hatched, they began eating at the brain until it drove the poor man insane.

Medically and entomologically, this is an impossibility. But, the myth stays with us. The Anglo-Saxons might have started the yarn. "Earwiga" means "ear creature."

Their metamorphosis is gradual. The nymphs are very similar to the adults except in size. They are ground nesters, and dig "caves" for their eggs. The female may lay up to 300 eggs (usually less). Then she does a strange thing. Like a wolf in its lair, she curls around the eggs and protects them from predators.

Many species have wings, and can fly over short distances. More often, however, they spread by the helping hand of man. Newspapers, flowers and plants, wood, and packages all carry earwigs to new homes.

There are several varieties found around homes. They may vary in size from less than half an inch to nearly 1½ inches. Coloring is usually reddish-brown to black, with certain species bearing light-colored stripes or spots. The tail length also varies from tiny, unobtrusive pincers to a huge, evil-looking pair. The pincers of the female are usually straighter than those of the male.

The forceps are used both in defense and while attacking other insects for food. Regardless of the opinion some people seem to have, the earwig is not poisonous. Like a scorpion, it raises its tail over its head. But, the two are not even distantly related. The pinch can be painful though.

Food is whatever they can chew. They have a strong liking for tender plants, and can do harm to gardens. The irregular holes they cause on leaves are quite similar to those made by slugs.

They also will devour many insects, including other earwigs. The exception is that they never seem to attack their own offspring.

When crushed, earwigs give off a bad odor. (Hence, the additional, incorrect name "stink bug.")

They find harborage wherever it is damp and dark, and are found in the same areas as crickets (although they rarely climb). An unfortunate habit of earwigs is that they love to crawl into piles of moist clothing, and sometimes into bedding.

Control begins with a cleaning outside of the house, and inside wherever dampness occurs. Diazinon in liquid, dust, or granules is applied to lawns, under boards and rocks, with a heavy concentration in the ever-present crack at the base of the house. Dust or spray crawl spaces.

Inside, treat all baseboards, with special attention to the basement, laundry rooms, and closets.

Fleas

When people speak of dangerous insects, the first mentioned are scorpions, black widows (neither of which are insects), and wasps. And yet, all these combined couldn't possibly equal the disease and death traced to the flea.

Because of their habit of feeding on rats, they are responsible for the spread of bubonic plague. As they jump from rat to rat, the disease soon infects a large portion of the rodent population. Then, when they turn to man for blood, the plague is transmitted to humans.

Although bubonic plague is now uncommon, cases do come up now and then. In June, 1976, a death in Arizona was traced to certain infected squirrels.

Other diseases fleas carry are typhus and tularemia. They are also known carriers of tapeworm.

The worm-like larvae do not feed on the host, but on organic matter found in the "nest." Because of their structure, they must have a moist area to survive. Any type of debris makes a good nest. Dust, dirt, bedding material, cracks in the floor, and sand also serve their purposes.

Fleas have a complete life cycle. This means that the larvae spin a cocoon where the pupa stays dormant until a host comes along. Any type of vibration, such as walking on the floor, will trigger the emergence of the adult flea from its cocoon. This is why a house which has had no host in it for some time, and is apparently free of fleas, will suddenly become quite active.

Like the larvae, the adults seek out dark, warm, moist spots. If exposed, they immediately try to burrow their way into any available crack, or into the dirt, sand, or carpet.

Control can be a long and exasperating process. Since they feed on any warm-blooded animal, any dogs, cats, etc. you have around, the home must be treated. This can be done by any competent veterinarian. He can also give you information on how to keep the animals clean of fleas.

Any suspected bedding material, yours and the animals', must be thoroughly cleaned or destroyed. Wash them in a strong soap, and then hang them in sunlight. (Don't just throw them in the trash bin. That's one way other people may get them.) Next, check outside for birds' nests, or the nests of mice and/or rats. Also look in the attic, chimney, in the basement, or under the house. All nests must be removed and destroyed.

Take some time to repair any holes where these animals might be coming in.

Inside, clean away any collected dust or debris. A normal vacuuming won't work. All cracks and crevices that have had years (in most cases) to collect dust and hair will have to be cleaned *completely*. Carpeting requires the same treatment. A good steam cleaning may do the trick nicely.

All this should be done before you apply chemicals. Fleas are rather tough. Insecticide alone may not be sufficient. It will be a hectic few days trying to get all this done. If just one part of it is done and then a period of time lapses before the next step, the fleas will have the chance to reinfest the cleaned areas.

There are several good residuals for fleas. Liquid concentrates seem to have the best results. Use either malathion, diazinon, or dursban.

Inside, treat as large an area as possible. (Read the label on the insecticide first.) Spray all cracks and crevices, concentrating your attention on the lower parts of the rooms. More often than not, it will be necessary to use some kind of flushing agent, such as pyrethrum, in a fine aerosol mist.

Outside, you have another big job ahead of you. If there is a crawl space beneath the home, it sometimes helps to spray it with water first. This will bring any fleas which have burrowed into the ground to the surface. Your residual treatment will then be more effective.

Again, on the outside you will be concentrating your efforts on the lower part of the house, and treating all cracks where the fleas might be harboring. Don't forget doghouses.

Using a hose attachment, treat as much of the yard as possible. The eggs may have dropped off of the animals on the lawn. By spraying just the perimeter, you will be taking care of only part of the problem area.

Attics should be treated with dust. All of the mentioned chemicals can be used, as can carbaryl. (Dust may also be used in the crawl space if desired.)

A word of caution. Since large areas, both inside and outside, are being treated, use more than a little precaution. Animals and children (and adults) will be coming into constant contact with the chemical. There is also an increased stain factor. So use common sense, and take your time.

Housefly

Flies

I've always found it strange how some people are reluctant to swat at a small housefly. "Oh, it's just a baby," they say.

The fly has a complete metamorphosis. This means that is goes from egg, to larva (maggot), to pupa, to adult. The small fly you see, and are reluctant to kill, is a full-grown adult. It won't grow any more.

Few insects are more exasperating. At times they seem to "have it in" for the human race. They display what seem to be emotions (they actually have none, as we know emotions). They'll begin a personal vendetta on one person, and become more determined the harder that person tries to shoo them away. If you swat at one and miss, he'll "dive bomb" you for the rest of the day. Or so it may seem.

Flies belong to the *Order* Diptera, which means that they have two wings instead of the four of most insects. Behind their wings are tiny balancing pads to assist them in flight.

As bad as fleas are, flies have caused more disease and death than any other insect. They feed and lay their eggs in any kind of organic waste. When they move on to the sugar bowl, or a plate of food, all the disease that has gathered in the wastes outside are brought along, and are then consumed by the human body. Diseases traced to flies include cholera, anthrax, dysentery, and typhoid.

The germs are carried in two ways. Most obvious is that while feeding on a diseased source, bacteria gathers inside the fly's body. Because of the way they feed, the disease is then transferred to the site of their next meal.

The other way, which is actually more important, is by means of the myriad of tiny hairs on their legs. The bacteria and micro-organisms cling to the hairs. Anything the fly touches can then become infected.

As mentioned, they feed in a somewhat peculiar fashion. A coiled proboscis, located just below the compound eyes on the head, is unwound and set against the food source. By a process of constant regurgitation, the food is brought up bit by bit. Each time the fly regurgitates, the food previously brought into the proboscis is set back against the food source.

There are many different species of fly. The one most people immediately think of is the common housefly. Other members of Diptera are the stable fly, horsefly, deerfly, fruit fly, drain fly, cluster fly, and others. Gnats and mosquitoes are also members of Diptera, but, because of their unique characteristics, will be covered in their own sections.

The female housefly can readily be distinguished from the male by pressing down on the back. On the female a short ovipositor will stick out. With this she lays her brood of eggs.

If it were possible for every egg to hatch, and for the young to develop unhindered, a single pair of flies may be responsible for 191 million million million flies. (191,000,000,000,000,000,000.) This is, fortunately, only a mathematical probability. Such a thing could not take place.

They feed on any fermenting organic matter, with a preference for pig and human waste. Horse and cow manure are also common feeding places. This waste matter is more ''popular'' if it is one day old. Other likely spots are wherever garbage is left.

The eggs are deposited in the same places; about 100 eggs every few days. The eggs will hatch in about 10 hours. The newborn maggot begins feeding immediately and continues to do so for about a week. It then leaves the moist food source and finds a dry spot, often by burying itself in the ground, or by crawling beneath any available rock or board.

The cocoon is made of the outer skin of the maggot. It goes through the transformation from larva to adult in anything from a few days to a month, depending on conditions. After a few hours it is ready to start life as the pesky buzz bomber.

If you figure it out, it's easy to see how flies become so numerous under the right conditions. The complete cycle from egg to egg-laying adult can take place in a week. A fly can become a great-great-grandmother in less than a month, and have literally millions of offspring.

Although their life span is usually only a few months, they have been known to survive for 1½ years. They spread both by being hauled around as maggots or pupae, and by flying. Generally they will not fly more than a mile from the original home site, but at times have been found to fly nearly thirty miles.

Control is difficult, and is more a matter of diligent sanitation than of chemical application. Corrals and stables must be kept clean on a daily (or more) basis. If the manure cannot be carted away, then spread it out as thinly as possible. If flies do lay their eggs in it, the combined drying power of the sun and wind will kill many of the larvae. Garbage *of any kind* should never be allowed to stand uncovered. (Organic manure used for gardens is often safe because it has aged and is not attractive to flies. But, if any other material is added, such as rotting fruits and vegetables in the compost piles, flies will most definitely find it.)

Treating compost or livestock areas is not such a good idea. Remember, flies breed quickly, and the chemicals would have to be applied every day. The animals would be in constant contact with it.

Immediately around the home there are a few things that can be done to somewhat decrease the numbers. There are fly traps; cages made of screen that act as a maze to trap the flies. If you care to make the investment, there are also electronic fly machines, most of which have an electrified grid to kill the flies. Automatic misting machines can also be purchased or rented from many pest control companies. Through a timing circuit, they release a measured amount of insecticide mist every 15 minutes. Pest strips may also be used. The trouble with these, and with the automatic misters is that even a slight wind will blow away the insecticide and render them all but useless. The pest strip also sometimes melts.

The best method is to resort to the old-fashioned strips of fly-paper. Not only is this the least expensive method, it will probably be the most effective.

Chemical control is usually rather ineffective, but may help when used along with proper sanitation. Areas where animals aren't kept may be sprayed with malathion or diazinon. Liquid formulations are best. Pieces of twine or cord may be dipped in either chemical, and when dry, can be hung or draped over ceiling beams. Caution should be exercised if this is done. Be sure that the ropes will not fall where a child or animal might chew on them. And, never use them around birds, such as chickens or pigeons.

To keep flies out of the house make sure that all screens are in good repair.

Drain flies, moth flies, fruit flies, and a few other closely related pests are sometimes found in and around the home. The famous Drosophilia (fruit fly) is often used by scientists and teachers because of its ability to propagate quickly. The others reproduce nearly as quickly, which means control can be difficult. The trick is to remove their sources of food.

Around the home, their food consists largely of fermenting waste found in drains, such as the scum that tends to build up over the years. They'll also go after fruit and vegetables. Clean the drains, and if possible keep them covered. Discard any rotting or overly ripe fruit, and keep all other fruit and vegetables stored away or at least covered.

In a few cases, the flies may be nesting inside a damp wall. Unless you have the proper equipment for drilling the wall, and for injecting the insecticide, it may be necessary to call in a professional who does.

Horseflies and deerflies are large, blood-sucking insects. Their bite can be quite painful. Control is the same as for houseflies.

Gnats

Gnats are also listed as Diptera, and are closely related to flies. Most are blood-sucking insects. Almost all are extremely irritating, with a habit of flying into the ears, eyes, and nose in search of moisture.

They have been known to transmit certain diseases. Eye gnats carry microorganisms to healthy eyes from diseased eyes and also may scratch the cornea, helping other bacteria develop.

In the western states, newcomers are sometimes surprised to hear tales of midnight attacks of ''no-see-ums.'' They are actually a type of gnat, very small and hard to see. They nest in marshes, or in holes of rotten trees. It's hard to seal the house well enough to keep them all out. Chemical treatment along windows, doors, and any tiny openings will help. (Use malathion or diazinon.) Sometimes spraying the lawn will help.

Clean up the surroundings to reduce the number of possible nests.

Gophers

''Try flooding them out with water.''

But, the water pours in for hours, only to come up hundreds of feet away. The next day, the gopher is back.

''Try sticking an exhaust pipe in the hole.''

Also, no results.

So, you try baits, both commercial and ''home remedies'' suggested by a neighbor. And, again negative results.

Brave souls might try cyanide (hydrocyanic acid, to be more accurate). If used near plants or bushes, the usual result is dead plants, and a gopher that has moved twenty feet to a new location. That, plus the danger involved.

I've even known people who have tried using bull snakes and gopher snakes. Forcing a snake to go where he doesn't want to is no easy task in itself. After you've accomplished it, chances are the snake will only come out another hole. Unless it's hungry, it probably won't bother the gopher. And, snakes don't eat very often.

Use of ferrets is also a solution that has been tried. Ferrets are altogether ruthless (but are gentle pets). But, like most animals that are undomesticated by nature, it may just run away.

Some dogs love to hunt gophers. They'll lay by a gopher hole for hours, and attack the second the gopher shows its head. The trouble here is that the dog may decide not to wait, and begin digging. A further, although relatively minor, danger is that the gopher may be carrying a disease (like rabies).

About the only way I have ever found successful is to use traps. The "Macabee" trap is probably the best, and can be purchased at many nurseries, hardware stores, and do-it-yourself stores. It is a strong wire contraption. Two prongs are held back by a wire from the trigger plate. When the gopher touches the plate, the prongs snap shut, stabbing the gopher in its sides.

Directions for setting the trap are usually with the package. Open a fresh hole and place the trap inside the tunnel. (A long spoon is an excellent tool for opening the holes.) A piece of twine tied to the trap, and then to a strong stick above ground will not only save many traps, but will also save you the problem of digging up the trap every day. If you make a loop in the twine you can easily see if the trap has been sprung or not. As soon as the gopher is stabbed, it will run back into the tunnel. This will straighten out the loop.

Finding the proper hole isn't as easy as it sounds. A fair-sized gopher of about 10 inches can dig almost 300 feet of tunnel per day. Many of these tunnels are soon blocked, either by nature or by the gopher. The main runs are used most often. Each time you smooth the mound over a main run and tamp it shut, the gopher digs it out again. You may have to "plant" the traps early in the morning or in the evening. Gophers are nocturnal.

Lacewings

The pretty green lacewing is not a pest, but a wonderful predator. Both the adult and the young (often called an aphid lion) feed on aphids and other plant pests.

The adult is up to ¾ inch in length; a soft, pale green with delicate lacy wings. If you look closely, you'll see that it has golden eyes.

Whenever possible, do not kill these. At times they may become somewhat of a nuisance around lights at night, but because of their beneficial nature, they are well worth the bother.

Lady Beetles

Most people know what a ladybug is, and know that they are good to have around. Like the lacewing, both the adult and the young are important predators of aphids and scale.

Their names comes to us from the Middle Ages. Because they are so beneficial, they were dedicated to the Virgin Mary, and called "Beetles of Our Lady." The common name of ladybug is inaccurate. They are a beetle, not a bug.

The one most people know is the variety that is red with black spots. They can also be black with red and yellow spots, yellow with black spots, and other combinations. (The number of spots does *not* indicate the age.)

Eggs are laid on the same plants where their food is. When it begins to get cold, the adults migrate to certain predictable locations. Firms which sell these beetles to farmers find these places and store the huge numbers of "hibernating" adults.

Lady beetles are predators, and are beneficial to gardens and lawns. An unacceptable infestation is almost a sure sign that there is an abundance of other pests to feed on. In this case, the treatment is the same as for other beetles. Be sure to find out what is attracting them, because this is the real problem.

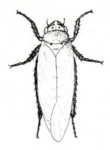

Leafhoppers

Leafhoppers are related to the true bugs. They are sap suckers, and cause considerable damage to leaves. The first signs of their presence is the discoloration and wilting of the plant. They are also known carriers of certain plant viruses.

The adults feed on the plant and lay their eggs in tiny slits made in the stems. This practice causes more damage than their feeding because it blocks the sap veins. After the eggs hatch, the nymphs fall off, and feed on decaying vegetation in the ground. Upon attaining adulthood, they return to the leaves to replay the cycle.

Although leafhoppers are tiny creatures, they are quite visible. If you take some time, and a magnifying glass, you will see that the shape of their bodies is like something out of a science-fiction program. Many look like futuristic space machines.

Control is best realized with constant vigilance. Diazinon, malathion, or carbaryl (dust or liquid) should be applied, starting in spring and early summer. Treat the stems, under leaves, and the ground around the plant. It may be advantageous to turn the ground a little before treatment, exposing the nymphs to the chemical.

Leaf Miners

Despite the similarity in name, leaf miners are not related to leafhoppers. Instead, many are related to flies (Diptera). Others are related to butterflies and moths (Lepidoptera), and still others to the ant group (Hymenoptera).

Oak Leaf Miner

Regardless of their relatives, they all do the same type of damage. The adult lays its eggs by piercing the leaf with its ovipositor. The larvae, when hatched, proceed to "mine" the leaf, leaving tunnels as they consume the plant and its juices. The damage is often seen as tiny light lines running just under the surface, or the leaves become splotchy and dried from lack of plant fluid. Some roll the edge of the leaf and feed within the fold.

Control is extremely difficult. Chemicals are all but useless, although frequent spraying with diazinon (dust or liquid) might help kill the adults as they lay the eggs, and the small number of larvae that come to the surface.

It is more effective, however, to simply pick the guilty leaves and burn them.

Lice

Like with bedbugs, the number of lice problems has decreased considerably. They've been around since the time of the caveman. It is believed that the body louse is a descendant of the vegetation-eating wood louse and the book louse. The louse has one unusual characteristic: its ability to take on the color of its host.

The myths and legends that have built up around lice are many and amazing. They have been used in the treatment of jaundice, and even for electing a mayor. (The first beard the louse crawled into was the man to be mayor.)

Unlike most insects, lice have no wings. Their legs are strong, each with a claw well suited to both fast crawling and clinging. You can become their host by contact with infested material or people. Consequently, they become quite a problem whenever crowds are together over a period of time.

The head louse prefers hair as a nesting spot. The body louse is thought to have evolved with the advent of clothing. The two can still interbreed.

The third type of louse that bothers man is one that many people joke about, but prefer not to talk about seriously. It is the pubic louse, sometimes called "crabs." As the name suggests, this louse somewhat resembles a crab, and prefers pubic areas on the body where hair grows. It will also infest the armpits, the eyebrows, and a beard or mustache.

Control is usually a matter for the family doctor or Health Department. About the only thing the homeowner can do is to try and find the source and eliminate it as much as possible. Bedding and clothing should be washed or burned. Frequent baths sometimes help with body or head lice, but should not be relied on by themselves.

Mantids

Like lacewings, mantids are not included as a pest but as a beneficial predator. Those who have seen them may find it hard to believe that they are related to cockroaches.

There are many different species. They range from about ½ inch to 6 inches. Color varies also, from tan or grey to a beautiful green.

The "walkingstick" looks just like his name. He moves very slowly, and may stand still for hours. He has a remarkable ability to take on the color of his background, which is his chief defense against the hungry jaws of birds and lizards. They are vegetarians,

Praying Mantis

and eat leaves. Although they are not a predator, neither are they a pest (except when they occur in large numbers).

The praying mantis is probably the most vicious, ruthless creature on earth. Watching them, you get the impression that they possess almost human characteristics. They swivel their triangular heads, watching any motions. If disturbed, they act as if they will tear off your finger. A few will stand up, and spreading their wings to make themselves appear more fearsome, hold their "hands" up as if to box. Despite this aggressive action, they will not bite. In fact, many people (especially the Chinese) keep them as pets.

Their food is anything that moves that isn't too large for them to handle. They eat all sorts of insects and spiders, and will also eat each other. Happily, they are the eternal enemies of black widow spiders. If both are placed in a cage, quite often a fight will take place, with the mantis (if she is large enough) the victor.

They sit patiently waiting for an insect to wander near. The front legs are poised like hands praying for a tasty meal. When "dinner" comes close enough they become a blur of speed, grasping the insect in those powerful hands, and immediately the mantis begins eating it. First the jaws pierce through the insect's hard exoskeleton, then the mantis devours the soft inside and drops the empty shell, and waits for the next meal.

Eggs are laid in a braided egg capsule. The case adheres to a branch or to the side of a building. Here they go through the winter and hatch out in the spring. As they have a gradual metamorphosis, the young mantids look very much like the adults except in size. They molt their outer skins as they grow.

If you are fortunate to have these around the house and garden, go out of your way to keep them alive.

Mealy Bugs

Mealy bugs cause large amounts of damage to almost every known plant. They are small, ¼ inch or less, and covered with a waxy white excretion. The body is somewhat oval.

They feed on plant juices, causing malformation, drying and wilting. Like aphids, ants often tend them for the honeydew they excrete. This honeydew may cause a buildup of a black, sooty fungus.

Control is much the same as for aphids. Be sure to spray or dust under the leaves, and down inside. Diazinon, malathion, or carbaryl may be used in dust or liquid form. As with aphids, apply the insecticide lightly, and often.

Mice

The old joke goes that women and elephants are scared of mice. Although elephants will, at times, spook at their scurrying movements, women usually feel only hatred and disgust when mice invade the house (men do, too).

Many fires in the home have been traced to mice chewing on electrical wiring. They will also chew holes in walls, furniture, and just about anything else you don't want them to.

The front incisor teeth grow more than an inch per year. If the mouse doesn't chew constantly, they will ''plug'' up his mouth and he will starve.

House Mouse

The name "mouse" comes from the oldest language in the world, Sanskrit. "Musha" in Sanskrit means "to steal," which is a rather apt name for a mouse. It scurries about and isn't adverse at all to taking small items into its nest.

The common house mouse is grey to light brown in color and weighs between ½ and ¾ of an ounce. Its tail is covered with fine hair (unlike the rat which has a fleshly tail) and is quite long, about the same length as the body of the mouse. They live just over one year, but in that short time they can produce as many as eight or more litters of six mice per litter.

The young are pitiful to look at. They are pink, bald, and obviously totally dependent on the mother. (The father might take to nibbling on them if he stays around.) In about 45 days after the birth, the mouse is fully mature and ready to become a parent.

A rat will eat an entire meal at one time. The mouse is more of a "picker." He eats a little, runs a little, eats a little, and runs again.

Their food consists of whatever humans eat, and a little more. To satisfy their instinct for chewing they may damage or completely consume wood, paper, lead, leather, and just about anything that's nearby.

They nest within a maximum of 25 feet of their food source. Only on extremely rare occasions will they wander further than this in search of food, once they have found a suitable home.

Besides the obvious damage of food, they also contaminate it with their feces. Their bite can carry several diseases, including tularemia. The mites and other parasites that feed on mice and rats spread even more dangerous diseases, such as plague.

All in all, they aren't very pleasant things to have around.

Any hole large enough to admit your little finger is like an open door to a mouse. And if the hole is too small, they merely chew it to the proper size. Mice also climb. A length of sheetrock is rough enough, not only for climbing, but also for stopping. Only polished metal and glass are slick enough to prevent their climbing.

Any holes, at *all* heights, should be tightly plugged. Plaster, sheet metal, or steel wool work well, and the mouse won't be able to chew its way through them. Check behind appliances for other holes. Entry points for pipes must also be blocked.

Food should be protected in tight containers. Paper and cardboard boxes aren't much of a deterrent. Metal cans, glass jars, or heavy plastic are better.

Once you have determined where they are nesting, feeding, and drinking, and once you've eliminated these, the mice might move away of their own accord.

If they don't, you can resort to two different methods: poisons or traps.

There are two forms of poison available to the homeowner. One is a premixed bait, usually made of some kind of grain. The active ingredient is usually warfarin, with a recommended strength of .025 percent. This might not seem like much, but remember, a mouse weighs less than an ounce. At this strength, it will take approximately body weight to kill. It is worth the extra wait to be certain that the poison won't kill one of your children or pets.

The second poison is harder to come by, and more dangerous to use. The chemical is fumarin, and is in a water-soluble powder. The use of both solid and liquid baits will ensure that both eating and drinking will further poison the mouse.

Of course, if the bait isn't placed properly, it will be useless. Mice run along the walls, and will eat the first thing they come to. Make sure that it's your bait. If you've found the nest, it's easy. If not, look for possible runways, and put the bait there, close against the walls.

Be absolutely sure that the bait is out of sight and out of reach of pets and children! Most commercial baits come in sealed boxes. If yours doesn't, be sure to put it in one. Not only is it safer, it is less messy.

There are two types of traps. The spring trap snaps shut on the mouse, crushing it. Live traps appeal to the mouse's natural curiosity. As it wanders into the opening, the trap closes.

Live traps are more expensive, and have to be emptied, but they have the advantage of being safer. Be careful not to just let the mouse loose again. Little fingers and paws won't be injured or broken by them. Also, most types of live trap will catch a number of mice before they have to be reset. Another word of caution if you use live traps. Even though they don't require frequent resetting, be sure to empty them, or the mouse will die and begin to smell. There are two companies that make live traps that you shouldn't have any trouble finding; Hav-a-hart and Ketch-all.

Spring traps are easy to get, and less expensive. If placed carefully, they can do the trick nicely, and safely. Remember where you've put them, and check them regularly. If you wish to bait them, use something soft, like bacon fat or butter, so the mouse will have to chew on the trigger.

Placement of traps is the same as for bait. Put them along runways, and close to the wall. The trigger of spring traps should be next to the wall. It will stand more chance of catching a mouse that way. After setting it, push it against the wall with a pencil. It might save your finger if the trigger is sensitive and the trap springs shut.

To make the trigger more sensitive, you can nail or glue a piece of metal or cardboard to it. The added weight will help, and the size of the trigger will be increased.

For years, professional exterminators have made use of another mouse catcher—the glue board. A special, very sticky glue is spread over a piece of wood or cardboard, and placed along a mouse run. Bait is sometimes used to better attract the mouse.

More recently, several companies have come out with a "public" version. Usually these are cardboard tubes with the glue inside the tunnel. A mouse that touches the sticky surface is stuck tight. Its struggles only serve to get it more tightly stuck.

The advantage of these is that they are clean, safe, and extremely reliable. Also, unlike glue boards, the premade units are not at all messy. The disadvantages are that they are costly, can only be used once, and are prone to losing their stickiness in dusty areas.

Some people also object to the way they work. The mouse is trapped in the glue and is stuck there, alive. It has to be killed or allowed to starve to death. For some it is also disconcerting when the mouse's struggles have dipped its nose to the glue, causing the mouse to suffocate.

Nevertheless, the disposable, premade glue trap is one of the most effective methods of catching mice. And, actually, they are no more cruel than any other kind of trap as long as you check them regularly.

Millipedes

Each segment of the millipede's body has two pairs of legs. In motion they appear to be like hundreds of fine hairs that "crush" the animal along. (It is obviously *not* an insect.)

Many people mistake millipedes for centipedes. The two are completely unalike, however. The centipede has broad, distinct segments. It's legs are always out from the body. The millipede is smooth in appearance and the legs are directly beneath the body. Another happy difference is that millipedes never bite, nor are they poisonous.

They can become a terrible pest, both in the garden and in the house. They eat both decaying and live plants. Roots and ground leaves can suffer damage. Rarely, they will invade the house, often in great numbers, and make pests of themselves by clambering into basements and climbing walls in the living room.

Millipedes are usually brown, or reddish-brown colored, and about 1½ inches in length. The first step to control them is to remove their food and nesting areas. Clean up the outside, remove debris, dead vegetation, rocks, boards, and any other objects they might hide under. Repair any water leaks and seal all openings into the house.

Granules spread around the perimeter are useful. (Diazinon is an excellent choice.) Using diazinon or carbaryl, spray nearby areas and the close perimeter. Crawl spaces should be sprayed or dusted with diazinon or Sevin.

Inside, use the same chemicals. Treat the doors, around pipes, and wherever else the millipedes might gain access. Moist cellars are favorite harborage points, and should also be sprayed.

MITES

Mites are so tiny, and of so many different variety, that people find it hard to believe that they are not insects. Spiders and scorpions, they can understand. At least *they* are large and totally different in appearance from insects. The mite is so small, that many people haven't seen what it looks like.

They are arachnids, as are spiders, scorpions, and ticks, have eight legs (instead of the six of an insect), and two body segments. All have piercing/sucking mouthparts. Many feed on plant sap, others on animal blood. This latter variety is responsible for the spread of several serious diseases.

Generally, the body is covered with fine hairs which act as the mite's sensory apparatus. The simple eyes, when present, serve only to detect light.

The garden species attach themselves to a variety of plants and drain them of life. Lawns are the most frequently affected plant life. Certain vegetables, fruits, bushes, and ornamental plants are damaged by other species.

Spider Mite

Spider mites are reddish in color and make themselves known by the silky webs they weave. Blister mites are famous invaders of fruit trees and leave dark blisters. Many species cause yellowing of the plant, or malformation.

Treatment for all plant mites is about the same. Ordinary water will help in many cases (such as with the spider mite). A good insecticide is diazinon. Liquid works in most cases; dust works better in others, depending on the plant, and environmental conditions.

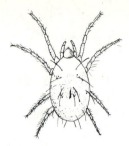

Spider Mite

Lawns almost always require repeated treatments. For those species that like to crawl into the house, you will also have to spray the perimeter, windows, and doors. Treat the affected area with a relatively heavy stream.

The blood-sucking mites are more of a problem. The diseases they can be carrying make them even more so. At best, they can be extremely irritating.

Chigger mites have the ability to release a toxin into the bloodstream as they feed. These sores can become inflamed, even infected. Quite often they originate on the lawn, or in nearby grasses or woods. Spray these areas somewhat heavily with diazinon (liquid), and repeat the treatment regularly.

This same treatment can control other mites with similar habits. For the mouse mite, rat mite, and others, you must rid yourself of the animals that help to spread them. (See the sections on these pests.) Spray all baseboards, cracks and crevices, damp areas, etc. with diazinon. Attics, crawl spaces, and any spot where an animal might make its nest should be sprayed or dusted. In cases of more serious infestation it may be advantageous to use a flushing chemical.

Mosquitoes

One of the most well-known insects is the mosquito. Most people know where they nest, what the larvae look like, what diseases they spread, and what sort of control is used.

Once again we come across a relative to the fly. The mosquito has a single pair of wings, and is, therefore, of the Diptera *Order*. Like certain gnats, horseflies, and deerflies, the mosquito is a bloodsucker. And, like the horsefly, it is only the female that feeds on blood. The male satisfies his hunger on plant sap. Surprisingly,

the process serves to pollinate certain plants that bees will not go near.

Eggs are laid in batches of up to 200 on the surface of any standing water, or in moist soil. In two or three days the eggs hatch. Certain species require periods of cold or dryness for the eggs to hatch. The larvae feed on tiny plant and animal life. Stagnant pools provide an abundance of this food.

The pupal stage is also spent in the water. Usually within two days the adult emerges.

The two most common diseases spread by mosquitoes are yellow fever and malaria. With modern vaccines, even these have ceased to be such a danger. The danger used to be very real, however, and took many lives. Some historians believe that this is one of the reasons ancient Rome fell victim to malaria (and plague.)

More recently, several cases of encephalomyelitis were traced to mosquitoes in Arizona. (Yes, there are mosquitoes even in the dry, desert regions.)

Control begins with a complete survey of your property. As mosquitoes rarely fly more than a mile from the nesting area, you should be able to determine very quickly if the problem is one for the city to handle.

Any standing water must be eliminated. Check under the house, if it has a crawl space. Don't overlook what may seem to be unlikely spots, such as cans and old tires. Gutters should also be checked.

Using malathion or carbaryl, spray bushes and shrubs. The underside of leaves is a favorite hiding spot during the day. Also spray the lawn under the house, porch, and any other area that might gather moisture or provide shade.

As with flies and gnats, check screens and openings around windows and doors.

Moths

Moths get their name from the German, "motte." (Others think it may have been derived from the Anglo-Saxon word, "mothe.") Both words mean, more or less, "mouth."

As with all flying insects, control of moths is difficult. The adult stage of many species do not eat. If they are of a damage-causing species, such as the clothes moth, by the time you see the adult, it is too late to do anything about him.

The clothes moth has a complete metamorphosis. It is the larvae that does the damage. As mentioned above, the adult cannot eat because of its underdeveloped mouthparts. The larvae eat not only cloth, but also fur, feathers, dead insects, and wool. Occasionally they will feed on living insects and also on grain.

They have a marvelous ability to hide. They avoid light, and will quickly disappear into a piece of clothing if exposed.

Eggs are most often placed between threads. If cloth dust gathers in crevices, they may also be found there. After about a week the egg hatches into a tiny caterpillar. Depending on conditions, the larva will weave its cocoon after a period of up to 2½ years. Three to six weeks later it emerges as the adult moth.

A thorough cleaning of the clothes and furs is a great help. Mothproofing should normally be left to professionals. A good dry cleaning firm may be of help in providing this service. If the infestation isn't too severe, mothballs, or a cedar-lined chest or closet will also help.

Wood Rat

Rats

    Having rats around has all the disadvantages of having mice, and then some. They are larger and stronger. The damage they do is usually more extensive. And, they are more vicious. Even when cornered, a mouse will only quiver its tail in fright. A rat could turn and attack.

    They are one of the few creatures that kill ''for the fun of it.'' (Two others are wolverines and man.) They are particularly fascinated by chickens. They might kill a chicken for sport and then leave it. Chicken eggs are then taken from the nests, then eaten or destroyed.

    Chickens aren't the only birds that are so bothered by rats. In the wild, rats attack, eat, and destroy all types of birds, and their precious eggs.

    Some rats are excellent at climbing. Birds roosting in trees aren't safe from their raids. Ground-nesting birds fare even worse.

    Like mice, the front incisor teeth of a rat grow at a tremendous rate. They must chew constantly if they are to survive. Mice, however, are more or less limited to soft articles, and rarely chew anything harder than wood. Rats have been known to chew holes through metal pipes.

    Mice generally eat grains. Rats will readily eat flesh, including living flesh. Rat bites are common occurrences in certain parts of the world. Entire toes and fingers have been eaten off by rats.

    If this weren't bad enough, rats carry rabies, several plagues, jaundice, and typhus (among others). Their contamination of food sources can also result in an outbreak of dysentery and food poisoning.

Obviously, rats aren't the most desirable creatures to have around.

The more common Norwegian rat (or brown rat) has coloring that varies from grey, to black, to auburn. This rat probably came to our country on ships carrying cargo from other parts of the world. It still does, for that matter. Another common name for it is the wharf rat, or sewer rat.

The snout is somewhat blunt. The tail is relatively short, being just a little shorter than the body. An adult reaches lengths of about 18 inches, and can weigh over a pound.

Another common rat is the roof rat, sometimes called the grey-bellied rat. Unlike the Norwegian rat, which prefers to stay on the ground, the roof rat lives up to its name by climbing to higher elevations. Actually, it is more common on ships than is the Norwegian rat.

The back is usually a brown, slowly changing to grey around the belly. Subspecies may be black with a white, or off-white, belly, or brown with a white belly. The snout is sharper in appearance than that of the Norwegian rat. The single-color tail is longer than the body.

Where the Norwegian rat and the roof rat are present, a state of war exists. Being larger, the Norwegian rat usually drives the smaller rat away. Not only will the Norwegian rat attack his relative when confronted, but will at times search him out.

The Norwegian rat will climb, and the roof rat does come down, and even burrows into the ground on occasion. The next is always where it will not be disturbed. Outside, it may be a burrow in the ground, in garbage, boxes, woodpiles, or brush. The burrow might be deep in spots, as much as four feet.

Inside, they can be found wherever there is a place for them. Roof rats find their homes near ceiling rafters, or in attics. Both rats might homestead under the house, or in cellars.

After a pregnancy of three weeks, a litter of up to 22 is produced. In just a few months these young rats will reach maturity and be ready to further populate. In this life of 1½ years (roof rat) to 5 years (Norwegian rat) as many as 12 litters per year can be conceived.

But, their life expectancy is usually much shorter (thank God!).

About one year is average for both species. Litters are usually seven or eight young, with about seven litters per year, a little better, but not much.

Man's efforts at controlling rats works to a degree. But, in the wild, nature provides other controls. Cats play only a minor part, and will rarely attack an adult rat. Owls, skunks, and certain reptiles help to keep the numbers down. In other parts of the world, ferrets and even boa constrictors (and other snakes, such as the cobra) are kept around to kill the rats.

Chances are you don't have a boa or a ferret as a family pet. What can you do to control the rats?

The first step, of course, is to remove any possible nesting sites, and the food the rats are feeding on. Proper sanitation is the best method of controlling rats.

"Rat proofing" a house can be a long and tedious process, but well worth it if rats are getting in. Patching with wood will work only when the rats are few in number and not very persistent. Cement or metal (like sheet metal) are much better, as the rat will have a hard time chewing through them.

Once you have eliminated the source, and have sealed your home from their entry, let your neighbor know that there are rats around (if he doesn't know already). The rats you have driven out could move to his yard.

Few commercial baits will kill a rat. At least not very quickly. When combined with the other steps, it will have a better action. As with mice, warfarin is the usual poison used today. Poisons such as arsenic and 1080 (a particularly potent chemical with no known antidote) are considered too dangerous to use under most conditions.

Fumarin may be dissolved in water for further poisoning. Whether you use warfarin or fumarin, *be sure to place the poisons safely!* These poisons are designed to act against warm-blooded animals. And, children are definitely warm-blooded.

Traps for rats are basically the same as those used for mice, only larger and stronger. Where the small spring trap used to kill mice might fracture a finger, the larger version for rats can easily break bones, even if it hits against a larger bone. Set them cautiously, and away from small, curious hands or paws. The results could be disastrous otherwise.

Traps and bait should be placed where the rat will come across them. Trails are usually easy to find. The large droppings, although unpleasant (and unsanitary), are a dead giveaway. In many instances there will be dark, greasy marks where the rat has traveled. Damaged materials may also give clues.

The rats may be quite a distance from the food, however. Try to follow any clues back to the nest. If it's safe to do so, and *only* if it's safe to do so, set traps or bait near the nesting site. Keep in mind at all times that the poisons and traps used for rat control are extremely dangerous for other animals (including humans).

If you find the nest, destroy it, but again, *only* if it's safe to do so. Rats can, and will, attack, sometimes without provocation. The only truly safe way to eliminate rats is to eliminate possible nesting sites before the rats move in.

One additional word of caution. Gas bombs used for gophers should not be used for rats. The gases are just as poisonous, but the conditions are quite different. A gopher burrows deep into the ground and digs tunnels. The gases move through those tunnels.

A rat's nest, on the other hand, is often better ventilated, which means that the gas has plenty of places to seep into the open air. This makes the gas not only ineffective, but very dangerous.

In addition, a rat's nest is quite often filled with flammable materials. The most common gas bombs for gophers ignite and burn. Using them in a rat's nest can cause a fire, one which can spread out of control.

Soft Scale

Scale

There are many different types of scale. All are harmful to plant life. They drain the plant of its sap, and occur in incredible numbers.

Because they are so small (always less than ¼ inch), individuals are sometimes hard to pick out.

They have a flat, oval shape. Once they sink their proboscis into the plant, they remain more or less stationary, and give the appearance that the plant is covered with tiny, white, powdery scales.

Eggs are laid in large masses. The larvae crawl on the plant, find a nice spot where they can pupate and grow. The newly-hatched larvae begin eating immediately.

Along with the obvious damage to the plant, the honeydew they excrete promotes the growth of molds and fungi that give the plant a dirty appearance.

Control is the same as for aphids and mealy bugs. For lawn scales, treat the entire yard at least once a week until the problem is under control. Diazinon, malathion, or carbaryl can be used.

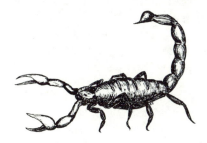

Scorpions

It's fairly easy to see why scientists believe that scorpions might have descended from the ocean lobsters. (Many inland areas were covered with water at one time.) The front claws are still present. The tail fin has been traded in for a thorn full of poison.

Many areas of the world have scorpions. Some are extremely dangerous, and death is almost a certainty after a sting. Scorpions found in this country (the greatest number are in the western states) aren't quite so deadly, but have been known to be fatal.

Oddly enough, it is the smaller species that are the most dangerous. The large varieties rely on size and strength for survival. Their sting does little more than poke holes. Even this can cause considerable discomfort, however.

A favorite hiding spot for scorpions is under objects, such as boards and rocks, especially where moisture is present. I have found the greatest number in, and around, desert washes. Debris caught against bushes makes wonderful homes.

Around the house they can be found in a number of places. They are good climbers, and can be found anywhere from ground to roof. Despite their adaptation to desert life, they hide from the hot sun, and prefer dark, secluded areas.

Stings usually result when the scorpion is disturbed. When a person accidentally puts his hand on one, or steps on it, it will almost certainly retaliate. Scorpions have short tempers. At times, something in his path (like your foot) is considered justifiable cause to sting.

Their desire to hide may drive them to seek shelter within clothing, or shoes. When the unfortunate owner puts them on, the scorpion is ever ready with his terrible thorn.

So fast is the strike, that it is estimated that a scorpion can sting 15 to 20 times in the time it requires to fall the length of the forearm. Fortunately, the poison runs out after only two or three stings.

If you are stung, immerse the wound in ice or cold water. The pain usually subsides quickly, but see a doctor. Don't take a chance.

Scorpions rarely sting while hunting for food. They seem to sense when its power is required, and save every drop of toxin for those occasions. Instead, they use the front claws both for capturing and holding their prey, and for tearing it into pieces small enough to eat.

Food consists of a wide assortment of things. The young have a particular liking for the soft-bodied termite, which is found in great abundance in the desert. When older, they eat cockroaches, small grasshoppers, and even pieces of beef and other garbage. (For those of you who would like to keep one, they survive quite well in captivity, and are easy to feed.)

The young are born live, and crawl onto the mother's back for the first few weeks of their existence. When large enough to fend for themselves, the mother will have nothing more to do with them.

Clean up the outside completely, but carefully. Exposed scorpions become particularly mean. Wear heavy gloves, or turn over objects with a stick. Kill any that you find. The smaller, more dangerous scorpions, can fit through a crack of 1/16 inch with no problem. So, be sure that all door and window seals are tight. (Remember, scorpions climb.) Any other holes should also be sealed. Check where pipes enter the house. Unless this maintenance is done, control will be ineffective.

Inside, you should also take a few precautions. Move beds away from the walls. As they cannot climb slick glass, the legs of the bed should be placed in jars. Keep clothes hung, and shake out shoes before putting them on. Scorpions move around at night, so if you get up to get a drink, don't walk barefoot.

They are hardy creatures, covered head to foot with a tough exoskeleton. Chemicals take quite a while to penetrate. Diazinon liquid residual should be applied to the entire perimeter and to all possible nesting sites. Dust well under the house and in attics with diazinon, Sevin, or chlordane.

Inside, spray all baseboards, windows, and doorways. Closets, laundry rooms, and basements are also favorite spots for scorpions and should be treated.

Silverfish

Entomologists believe that the silverfish is even older than the cockroach. They are usually white to grey in color, and dusty in appearance, with long antennae and usually three "tails."

Man is not the only creature they bother. They will also invade the nests of ants and termites. While in the ants' nest, the silverfish will steal the precious honeydew, and then run off before the ants can revenge their loss. In a termite colony the silverfish eats both the eggs and the young termites.

In the home he will feed on anything starchy. Silverfish are responsible for the destruction of many valuable documents, papers, and books. Animal and vegetable products, flour, and paper make up their diet. New homes are often overrun by silverfish that have come in on plaster board and other building material.

Normally it takes some time for damage to show itself. But, they have been known to eat large holes in all kinds of paper and certain cloth goods.

Throughout their life they do not consume liquid. When found in sinks it is because they have stumbled into them. As they cannot climb a slick surface they become trapped.

A good trap for them is a small jar with a small amount of flour in it. When something rough (such as masking tape) is placed around the outside, the silverfish can climb into the jar, but will be unable to get back out.

Their metamorphosis is gradual. Eggs are dropped wherever the female happens to be at the time. There are approximately 100 eggs per brood, with several broods per year. Their life span can be up to 3½ years.

Firebrats, which are found outside in western states, can be distinguished from other silverfish by the dark stripes along their bodies. They live under rocks, boards, or in cracks in wood.

Malathion and diazinon are the most effective chemicals and may be used either as liquid or as a dust. Treat along baseboards, in cracks, and in closets. Dust in attics and in crawl spaces or basements.

Outside, you will find very few silverfish except for firebrats. Treat around the perimeter with dust or liquid as well as around any possible nesting areas.

Snails and Slugs

The biggest problems with snails and slugs is the damage they can cause to plants. For the most part, they will be found only in wet areas. Snails can survive dryness for extended periods of time (up to 4 years) because of their ability to close their shell.

Eggs are laid in the same places that they live and feed. Slugs lay them in gelatinous masses of about 25 eggs. Snails usually dig a hole in the ground and plant their future brood of up to 200. Life expectancy can be as much as three years. If a snail has gone dormant because of a dry spell, this period may be extended.

Snail

Slug

Clean away as much decaying vegetation as you can. By eliminating their food source, you will automatically cut their numbers. Also, find the nests, and potential nests, and destroy them. Remove rocks and boards when possible. If there is a water leak, inside or out, repair it.

Because of their wet nature, granules are usually most effective, however, dust, liquid, or granules, almost any chemical will do. Dry crawl spaces may be treated with dust, liquid, or granules. About the only place *in* the house you'll find them is in very damp basements. Liquid residuals can be applied.

Commercial baits will probably provide better success. Some of these contain dangerous chemicals, however, and should be placed with great care. Be absolutely certain that children and pets won't get at the bait. The usual active ingredient is metaldehyde.

In placing either insecticides or bait, it is best to use them along the paths that the snails or slugs follow. Normally they will always travel the same paths to and from the nest.

Sowbugs and Pillbugs

Sowbugs and pillbugs, these armadillo-like creatures are not insects at all, but are crustaceans, related to the crayfish. To tell pillbugs from sowbugs is simple. The pillbug can roll itself into a tight ball. The sowbug can't, and has two appendages at the back of its body.

Sowbug

Pillbug

Both range from grey to black. They make their homes under objects, or in the ground. Food consists of decaying vegetation or small living plants, which makes them something of a garden pest. Occasionally, they stumble into the house, particularly into damp basements. They do no damage in the home, but can be irritating merely by their presence.

About twice a year, the female releases a brood of 25 white offspring. The young shed their first skin within a day, and subsequent skins about every two weeks until they have undergone eleven molts. Life expectancy is usually about two years.

Granules are usually best for control because of their damp nesting habits, with diazinon being one of the better chemicals. Spread the granules around any areas where there is vegetation, and under any boards or rocks. Using liquid diazinon, spray along the doorways to keep them from coming inside. If you have a crawl space, granulate or spray with diazinon, or, if the area is relatively dry, dust with carbaryl. (Remember, dust clumps if wetted.)

Inside, spray along doors and in the basement.

Spiders (General)

All spiders are poisonous, but only a very few are dangerous to man. In fact, spiders are beneficial. All are predatory, and kill and eat insects.

Occasionally, however, they become so numerous that they become a nuisance. And, some species, such as the black widow, have a dangerous bite.

Spiders live wherever there are insects. In size, they range from the microscopic to huge beasts up to 10 inches across. Colors are

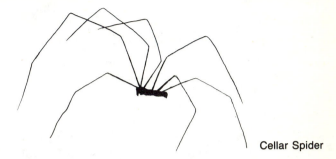

Cellar Spider

usually the tans and browns of earth, but may be bright or metallic colors.

Most make webs in an infinite variety of patterns. Others live in holes in the ground, under rocks, or even in underwater homes.

Control varies on the nesting habits of the problem spider, but always begins with a good housecleaning inside and out. Remove any possible nesting sites, at least those around the house itself. Repair any faulty seals around windows and doors, along with any other cracks or holes.

Chemical control of other insects will automatically cut down the spider population.

Diazinon is effective in liquid or dust. Carbaryl is also good in dust form. Apply these along overhangs, in woodpiles, crawl spaces, attics, and any other possible nesting spots. For lawn-nesting spiders, spray or granulate thoroughly.

Treat windows, doors, and cracks inside. Many spiders like cool, damp basements, and these areas will require special attention.

Spittlebugs

The adult spittlebug is similar in appearance to the leafhopper, but does no harm other than laying eggs. It is the nymphs that do the damage.

A sure sign of spittlebugs is the presence of foamy white spittle which gives them their name. The nymphs can be found under the froth, feeding on the plant's juices. The nymphs are just under ¼ inch in length and of a pink or yellow-green color.

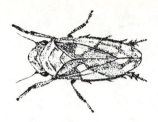

Normally they do not bother vegetables, but are definite pests of ornamental plants and many fruits. Other types will attack full-grown trees, especially the pine.

Clear away the "suds," and as many of the nymphs as possible, and destroy them. Then treat the affected plant with dust or liquid. Retreat as necessary. Diazinon gives good results.

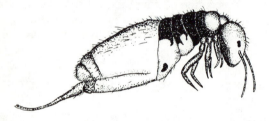

Springtails

The tiny springtails are sometimes mistaken for fleas because of their ability to jump. For this jumping they use a powerful "tail," which gives them their well-deserved name.

They can be found wherever it is dark and wet. Basements can suddenly become infested with them, much to the dismay of the homeowner. If there are no plants around, the best bet is to look in the drains.

Outside, they prefer moist ground, such as in gardens. Some varieties feed on plants in the garden and eat little holes in the leaves. Other types eat microscopic plant life and decaying vegetation.

In the basement, check drains and pipes, and under the insulation. If there are leaks, make the necessary repairs, and take whatever steps you can to reduce moisture. Using either malathion or diazinon, treat along all baseboards, in cracks and crevices, under sinks, and on pipes.

If you can determine the point of origin of the infestation, control will be much easier.

The garden springtail is usually purple in color. Look under leaves, wood chips, and moist areas. Treat along the perimeter and in the crack where the ground meets the house. Liquid residual or granules will give you the best results.

Tarantulas

Few spiders have received more attention by movie makers than the hairy tarantula. He is reputed to be instantly deadly, and extremely vicious. Neither is true.

With their powerful fangs they can inflict a somewhat painful bite, and may even draw blood. But, there is no danger from their poison.

Not only are they *not* vicious, but they are actually quite gentle. In fact, they make excellent pets and live up to 25 years. Even the wild tarantulas are docile, and will bite only when provoked.

They make their homes in holes or under rocks. At night, they prowl around for their food, which is usually larger insects such as grasshoppers. Their walk is slow, almost nonchalant. But, when their prey comes near they can move with blinding speed. Occasionally they will leap a few inches. (A common tale is of a tarantula leaping 10 or 20 feet and killing a horse with a single bite.)

If you are bothered by them (which is a rare occurrence) look around the yard for any holes. Rocks, logs, or woodpiles can also harbor them. The best way is simply to move or destroy the nest. The spider will be more than happy to move on.

If the problem persists, or if a nesting site can't feasibly be removed, treat the area with diazinon, malathion, or carbaryl. In most cases, dust will provide the best results.

Because of their size, they require a fairly large opening to enter the house. Look for any opening and repair it. A tarantula can climb, so don't forget to check attics and roofs.

Crawl spaces and attics should be dusted, especially if there is already an infestation.

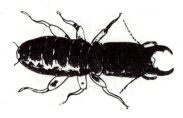

Termites

A great many people make the mistake of thinking that termites and ants are related. They both have a highly-developed social structure (as do bees), complete with workers, soldiers, and the all-important queen. But there the similarity stops.

In fact, ants and termites are deadly enemies, with ants usually coming out as the victors. Ants stage frequent raids on termite colonies, capturing the termites for food. Immediately the soldier termites jam their huge mandibles in the opening to the colony. While they are fearlessly defending the nest, the workers are busy sealing up the tunnels.

Even if the attack is repelled, the soldiers are doomed. Being separated from the life-giving humidity of the nest, they soon die.

The name termite was given to us by the Romans. "Termes," the Roman title for these insects, means "wood worm." (The worker termite *does* resemble a worm.)

It is estimated that there are more than 1,800 species of termite. Of this number, there are only about 50 species in the United States. The vast majority make their homes in tropical regions. Those species in America are native to it. Because of their particular requirements, termites are never carried in from foreign countries as are many other insects and rodents.

Without termites, our earth would be a very undesirable place to live. Large, hardy plants, such as trees, that have died, take a considerable length of time to decompose. The little termites make short work of it.

To digest their food, termites have an abundance of certain protozoa and amoebae (one-celled creatures). In some, fungi serves to break down the hard cellulose of the wood. A few do not eat wood, but molds and fungi. Many species have a combination of things working for them. When external fungi attacks the wood, breaking down the cellulose, the termites move in and consume the by-product.

The caste system is remarkable. When all the specialized classes of the colony work together, it is like a single, intelligent being. Like the cells of our bodies, they each perform a specific function.

Most numerous are the workers. They are small, white to grey creatures. These are the ones which do the actual damage. With their scissor-like jaws they clip minute pieces from the wood.

Unlike the ants, where the workers are sterile females, termite workers may be either male or female. They not only provide food for the colony, but also perform all other forms of menial labor. They build and repair tunnels, cultivate the fungi gardens, and feed the queen, king, soldiers, and the young.

The soldiers have developed huge mandibles for defense of the nest. In some species, workers are not present. All work is carried on by the nymphs or the soldiers.

The king and queen are responsible for the continuation of the colony. They begin as winged reproductives. If a male and female are lucky enough to find each other, and a suitable site before they are killed by the sun or predators, they begin to reproduce. Until the young are sufficiently developed to perform the work of the nest, the king and queen feed and care for the young, and keep the nest clean.

Once mated, the two stay together for life and produce some 3,153,600 eggs per year. Occasionally, there are secondary kings and queens in the same colony, which serve to reproduce on a minor scale. Their main function is to take over should the queen be killed.

There are three main types of termites: the subterranean, the dry-wood, and the damp-wood. Their names describe how each lives.

The subterranean finds its home underground. Colonies have been found as far down as 50 to 60 feet. From there, they tunnel their way up and into the food source. No wood is completely safe, but the heart of redwood doesn't entice them quite so much as pine or other soft woods.

Subterranean termites leave mud tubes. These tubes serve as a passageway for the termites. Inside, the termites carry the atmosphere of the nest. Without it they will die. Tubes are constructed whenever the worker finds it necessary to leave the protection of ground or wood, such as long cement piers. They may also be seen coming down from the ceiling like stalactites. New tubes can be built at a rate of several inches a day. If it becomes necessary to repair a broken tunnel, the job is done much faster (2 to 3 feet per day).

The dry-wood termite is found inside the wood at higher elevations. They feed on dry, seasoned wood by "drilling" into it. They need no contact with the ground, and spread by flying. Control is a matter for professional exterminators. The homeowner simply cannot effectively treat them.

The damp-wood termites are similar to the dry-woods in that they require no direct contact with the ground. They feed on moist and decaying wood. Once again, control is usually a matter for professionals.

Flower Thrip

Thrips

Although thrips are usually citrus pests, certain species will attack other fruits, as well as vegetables, grains, and flowers.

They are extremely tiny winged insects, often only ¹⁄₂₅ of an inch. They feed on the sap of plants, causing them to malform, or the buds to fail to open. They mar the leaves with a dotted, mottled appearance, along with tiny brown flecks of excrement.

Their coloring is usually yellow or light brown.

Treat the affected plants with diazinon, either dust or liquid. Retreatment will almost definitely be needed.

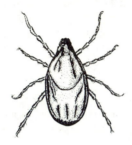

Ticks

Few pests are more hated than ticks. Their whole purpose in life seems to be in making every warm-blooded animal miserable. Not only do ticks drink the blood of their host, leaving it devitalized, but they also carry many serious diseases.

The adult tick has eight legs, which allows it membership in the order of Arachnidae. Along with their devitalizing and disease-carrying characteristics, ticks also can inject a type of toxin which, by itself, can paralyze the host.

Both the male and female feed on blood. Usually it is the female that makes her presence known as she swells to outrageous sizes. Most species require this gorging to complete the egg-laying cycle.

The brown dog tick (the tick is brown, not necessarily the dog) is perhaps the species most frequently encountered. It has a knack for finding blood vessels on animals. Between the toes, in the ears, and on the neck by the jugular veins, are favorite spots.

The male and female are about ¹⁄₁₆ inch before feeding. When gorged, the female may distend up to ½ inch. At this point, her color has changed from the normal deep brown to an ugly grey. She will then withdraw her beak and leave the host. After a suitable hiding place is found she drops her load of eggs. There may be as many as 3,000 eggs per lay.

After a few weeks to two months, the eggs hatch into tiny, six-legged nymphs. As soon as they can, these young will feed, and turn a dark blue in color. After two feedings, each followed by a molting, the tick is ready to assume adult life.

The brown dog tick can be found wherever there are dogs. Any crack or surface serves as a home and hiding place. They are good climbers. Don't be surprised to find them in cracks on or near the ceiling. Even if the dog is kept outside, the ticks have an unfortunate habit of working their way into the home.

To control them you must treat both the house and the dog(s). This way you'll be hitting the ticks both where they live and where they feed. If one part is done without the other, the chances are you've just wasted your time.

For the home, use diazinon, dursban, or malathion, preferably in liquid form. (Dusts tend to be difficult to use for ticks.) Spray the entire yard, with extra attention on spots where the dog spends the greatest amount of time. The holes and cracks in walls and fences make excellent hiding spots. And, if the neighbor's dog has ticks, they'll invade your yard by climbing over or through the fence.

As mentioned, ticks climb, so don't forget higher spots. Check along the eves and any cracks near roofs. The thousands of holes in brick and slump block walls may be harboring ticks. This may give the wall a spotted appearance. If the ticks are here, spray the entire wall.

Storerooms, woodpiles, boxes, and crawl spaces might also have ticks hiding in them. Clean the area as much as possible to lessen the number of possible homes. Then spray everyplace where the ticks might be. If the area is closed in, or if a crack is unusually deep, you might want to use pyrethrum to flush the ticks out across the residual you've already put down.

Inside, spray all baseboards, cracks, and crevices. Check in closets, clothing (especially stored clothing), furniture, and curtains. Ticks aren't too fussy as to whether their home is made of wood, cement, or cloth.

If the problem inside is serious, use a flushing agent. It maybe necessary to completely fog the house. Should this be necessary, take all of the normal precautions. Remove all animals from the house. Open food should be taken out or put in the refrigerator (if the seal is good). Cover all dishes and utensils. Finally, shut off any flames.

If the ticks on the dog aren't too bad, pull them off. Either with your fingers or tweezers, grasp the tick as close to the wound as possible. Try not to squeeze the body as this might force toxins into the dog. Twist the tick to prevent breaking off its head. Then, kill the tick. You can drop the tick into a bottle of alcohol or burn it.

Most drug and pet stores carry different products for treating the dog at home. These will work fine if the problem hasn't gotten out of hand, and for future retreatments. If there is a large number of ticks, it is usually safer to have a competent veterinarian "dip" the dog. Again, this should be done at the same time the house and yard are being treated.

Any bedding the dog has used must be destroyed, preferably by burning. The eggs are so numerous, and so tiny, that the best washing may not get them all out. Spraying the bedding is risky (the dog will be in constant contact with it), and probably won't work very well.

If your dog is allowed to wander, or if other dogs can come visiting, expect the problem to return. Tick-free yards can become infested by another dog just coming near the fence. And, all your dog has to do is step on another lawn that has ticks to pick them up again.

Try to contain the dog, and keep others away. Or, resign yourself to the task of constant vigilance. Retreatments won't be too difficult if you can keep the ticks to a minimum.

Other species of tick, such as fowl ticks, groundhog ticks, and fever ticks require the same sort of treatment. Instead of a dog as the host, these are brought in on birds or rodents. Chemical treatment alone won't do much good. You'll have to control the two- and four-legged carriers before any substantial control of the ticks can be realized.

True Bugs

A bug is an insect, but an insect isn't necessarily a bug. At least not in the true sense. The true bugs belong to the class of Hemiptera. They have piercing mouthparts and may be either plant feeders or bloodsuckers.

Chinch bug

The box elder is a true bug. Since most people have, at one time or another, seen a box elder, it serves as a good example of what a bug looks like. Most are dull in color, but a few have bright markings.

The chinch bug is an important garden pest. The young nymphs attack lawns, leaving large dead spots. Chinch bugs can grow to about ½ inch in length, and have dark red and white markings on a brown body. The eggs are laid on the roots and lower parts of the grass.

At the first sign of damage, spray the lawn with diazinon, malathion, or carbaryl. Retreatments will probably be necessary. If you start your treatments in the spring, the task will be much easier. After you've gained control, retreat regularly to keep the problem from coming back.

The squash bug is also a garden pest. By draining leaves of their juices, the plants begin to wilt and die. When crushed, this bug gives off a foul odor.

The famous stink bug has small glands on its ¾ inch body which emit the smell that gave it its name. They vary in color from brown to green to a shiny, gunmetal black with red marks. Some burrow into the ground. All feed on plants, such as tomatoes and other garden vegetables.

Control for the squash bug and the stink bug is essentially the same. Treat the infested area with malathion or diazinon. Either liquid or dust will work. Be sure to spray (or dust) under the leaves of the plant. Also spray or granulate the ground around the plants.

The assassin bug is fairly common, especially in warmer states. It is both carnivorous and a bloodsucker. Its bite can be extremely painful. And, because it may be found in the dens of pack rats, feeding on their blood, it has been known to transmit disease. The sore caused by the bite tends to infect for the same reason.

They favor woodpiles as a home, where they attack and eat other insects. (In this way they serve as a beneficial predator.) Control begins by clearing away as much of this as possible, and by ridding the grounds of rats. Then, spray the lawn and suspected nesting sites with diazinon or malathion. In some cases, dust works better than liquid.

Spray or dust in attics and crawl spaces, and any cracks. Also check storerooms and fences.

Holes and cracks leading into the home must be sealed to prevent the assassin bug from entering the house. (Their bite isn't the most pleasant way to wake up.) Windows, doors, and vents should be sprayed with diazinon or malathion.

Wasps

Like bees and ants, wasps belong to the order of Hymenoptera. A few are social, and build multicelled nests. Most species are loners, and make a cell for laying a single egg, after which they abandon the young.

Their sting is not barbed, like that of the bee, and won't stick in the skin. By stinging, the wasp need not give up his life. Along with this, the wasp often needs little provocation to sting.

Even when a colony is developed, it lasts only one year. Once the young grow into adulthood, they leave their home and find new places. The adults die off.

One of the most often encountered wasp is the yellow jacket. It builds its nest in the hollow of trees, in low bushes, or even in the ground. Like so many other wasps, it is attracted to anything sweet. It will also feed on other insects, which it chews into tiny pieces and feeds to the young.

The paper-nest wasp makes a comb-like nest of many cells out of saliva and chewed up vegetation and mud. Often there is a slender "stick" which holds the nest to overhangs.

The mud-dauber makes its nest out of layers of mud and saliva. Each cell is made separately, and a single egg is deposited inside. The end is then sealed over, and the adult goes away. In some species, the cells have a tubular shape; in others they are made up from concentric rings, each carefully molded and patted to the proper shape. A stung caterpillar or other insect is placed in the cell for the emerging larva to feed on.

Beneficial wasps include the cicada killer, the sand wasp (which feeds on flies), and the horse guard wasp (which also feeds on flies). In addition to their predatory benefits, certain types of inks and dyes are manufactured from the tannic acid found in some species of wasp, such as the gall wasp.

As with bees, extreme caution should be used in controlling wasps. To be effective, the insecticides must be shot directly into the nest. This might chase the wasps out, and into an attack.

A pyrethrum aerosol may first be used to stun the adults. If your sprayer has a stream setting, use this. It will keep you at a safer distance, and still deliver the chemical into the nest.

The best time to treat a nest is at night when the wasps are inside, and relatively inactive. Even then, be careful. A single wasp sting is bad enough. Don't put yourself in a position where you might receive multiple stings.

Diazinon or malathion are the best residuals. When feasible (and safe) dusts give better results. Even if the chemical doesn't get inside, the adult wasps will carry it inside on their feet.

After you're absolutely sure that there are no more adult wasps around or in the nest, destroy it and all larvae inside.

Use of fire is rarely a good idea. If the nest is near the house, the reasons are obvious. In a tree or bush the fire not only stands the chance of spreading, or getting out of control (even with the most careful preparations, an unexpected wind could carry a spark); the plant itself could be damaged.

Whiteflies

The garden seems to have more than its share of pests. Many of these pests are tiny, and breed so rapidly that control is next to impossible.

The whitefly is no exception. Both the white winged adult and the scale-like young feed on the precious plant sap, causing spots and yellowing. If the infestation is heavy, or if other pests are plaguing the garden, the plant could die.

When disturbed, the adults fly from the plant, like flakes of erratic white dust. The young attach themselves under the leaves and do not move at all. The honeydew they secrete causes the same black mold to form as the honeydew of aphids and true scale insects.

To control, lightly spray or dust the plant with diazinon. Be sure to hit under the leaves to kill the young. Contact as many insects directly as you can for a better kill. Retreatment will almost certainly be needed.

"Worms" and Caterpillars

The tent caterpillar and fall armyworm aren't worms at all, but are the caterpillar (larvae) of any of several species of insect. They attack lawns, gardens, fruits, and trees, and seem to defy all efforts of control.

Most feed on all parts of the plant. Others burrow into the fruit or bud. Damage can be as simple as small holes eaten out of a leaf, or it can be as extensive as the destruction of the entire plant.

Most often they make their home in the ground and crawl out in the evening when it is cooler. Others may hide in the foliage, or with the plant itself.

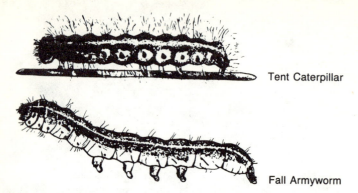

Tent Caterpillar

Fall Armyworm

Cutworms have the terrible habit of clipping the plant off at its base. These "worms" may be spotted or striped, and vary in size from about an inch to just over two inches.

Armyworms are usually found in southern states, but do migrate north. The newborn larvae are a grey-green, which slowly changes to almost a black in some species, and often have light-colored stripes. The southern armyworm is also called the laceworm, because of the pattern it leaves as it eats.

The caterpillar of the miller moth, so common in the west, feeds mainly on lawns. They are light brown in color and are covered with fine hairs. The pupae stage is often spent in drying dirt clumps.

Codling moths are fond of apples. Their eggs are laid in the developing apple. When the egg hatches, a small, usually pinkish caterpillar emerges and begins to feed on the fruit. Large portions of apple and pear harvests are damaged by these pests every year. (Almost everyone has had the experience of biting into an apple only to have their appetite destroyed by the presence of a tiny "worm.")

What you do for control depends on where you're finding the caterpillars, and what they're feeding on. Lawn-infesting worms are difficult. Usually the entire lawn must be treated repeatedly. Because the adults fly, the worms will probably come back again.

Caterpillars that climb will require either dust or liquid residual applied to the entire plant—carefully applied! Turn the soil around the plant and treat with liquid or granules. Once again, repeat the treatment as needed.

Diazinon, malathion, and carbaryl are all good poisons for caterpillars. If you have to treat vegetables or fruits, dust is often the preferred formulation because it has less tendency to absorb into the plant. In most cases, you can use one of the desiccant chemicals (silica gel or boric acid) when treating edible plants.

The adults are attracted to light. Gardens close to the house can become quite infested. Anti-bug lights will help, as will light-proof shades.

CONCLUSION

There are too many different pests to list them all. Fortunately, this isn't necessary. Treatment is similar for all pests with particular characteristics. For example, treatment for any kind of small flying insect is much the same; treatment for rodents of all kinds is similar.

If the pest you're facing isn't listed in this book, there are two steps you can take, both of which are tied together. One is to find another pest with similar characteristics that *is* listed. This will give you some clue as to how to handle the problem. Next, pay attention to what the pest is doing. Does it fly or walk? Where does it seem to be coming from? Where does it gather, and what is it eating? Is there anything you can do to break the reproduction cycle? (Perhaps controlling the adult is impossible, but stopping the pest at the larval stage is easy. Eliminate the larvae and you eliminate the adults, which stops the egg laying and prevents the growth of any other larvae.)

Use your powers of observation and your common sense to learn as much about the habits of the pest as you can. You don't have to be an expert or a scientist. Your goal is simply to watch and see what you can.

Apply the few basic rules that are the same for all pests that are covered throughout the book. Most all creatures have four primary needs—food, water, shelter, and entrance. Without all four of these, there are no pests for you to battle. Your goal is to eliminate, or reduce, all four basic factors. (This also serves to break the chain of reproduction.)

The first step is always proper cleanliness and household maintenance, especially in the immediate area where the pest is found. Remove all sources of food, water, and shelter. Without these necessities, the pests won't last long even if you do nothing else.

Keep in mind that cleaning just what is readily visible isn't enough. If you have roaches, for example, merely wiping off the counter won't help at all if there are other sources of food around. Sealing one hole and leaving six others still open (because you can't see them behind the stove) won't do much of anything.

Your next step is to find the source of the problem and eliminate it, if possible. Trapping one mouse as it runs behind the refrigerator won't help much if there are 20 generations of mice breeding happily in the attic. Swatting a flying insect won't cause much of a dent in the pest population that floods your back patio every night. (In fact, nightly bombings won't do much but provide temporary relief.) To gain any control, you have to find, and eliminate, the source.

For pests that enter from the outside, the key is to prevent, or at least slow, that entrance. This means that all windows, doors, vents, and walls have to be sealed. This helps to eliminate the fourth necessity—entrance.

To further assure that pests don't get in, check packages and bags. A number of pests can find their way into the home with the kind, although unwitting, help of a human (or animal). Isolating and putting new items into temporary "quarantine," such as using some kind of secure and sealed container, can help prevent an overall infestation from one buggy package of flour or cereal.

Application of insecticides is also a matter of common sense. For normal use, one of the usual residual chemicals will do the job just fine. For heavier infestations, or for problems where the pests are hiding, you may need to use a flushing agent, such as pyrethrum, with the residual applied first. Other problems may require a more thorough use of a flushing agent, up to, and including, "bombing" the entire house. (This should not be standard procedure, because of the effort involved for preparation and the potential dangers.)

One of the best all-around insecticides presently available is diazinon. It has a fair quick-kill capability, a good residual effect, and is still fairly safe to handle. Malathion is even safer, but is slower, more prone to breaking down under heat and sunlight, and has a more powerful odor.

Pyrethrins are the standard for knock-down or flushing. In aerosols they can be used to "fog" an entire house, if need be, and are safe if handled according to the label.

Liquid concentrates are generally best for most uses and are versatile. Dust tends to have a longer life and fewer harmful effects, especially on plants. Granules are useful only when there is moisture to activate them. All forms are potentially dangerous if misused. And all are actually less effective if applied too heavily, or when not mixed correctly. If the concentrate calls for 1 ounce per gallon, you're *not* going to do better by putting in 2 ounces.

The key is common sense. If your kitchen is infested, it doesn't make much sense to spray just the basement. Nor does it make much sense to complain about being infested with mice, rats, and roaches while your kitchen is a filthy grease pit, your yard looks like the city dump, and your home is "air conditioned" by numerous holes.

7

Conclusion

If you follow the directions and precautions of this book, and read the labels on all chemicals you use, you should have no problems. Pest control *can* be safe. All that is required is a little common sense.

However, should accidental poisoning take place, the speed with which you react is important. If the pesticide has been swallowed, or if contact was made with the concentrate, quickness is even more urgent.

First, remain calm. Stop and think. If there are two of you there besides the victim, send one immediately to call a doctor. And, make sure that the doctor is given all available information; name of the chemical, how the poisoning happened (if it was swallowed, spilled on the skin, etc.), approximate age and weight of the victim, and so on. The more he knows, the better he'll be able to administer help.

Unless the victim is in a place where he is still in contact with the poison, don't waste time moving him. Washing off the poison is more important. The longer the poison stays on the skin, the more poison will be absorbed. Use soap and water. (More information is in Chapter 2.)

If the person has stopped breathing, apply artificial respiration. To do this, make sure the victim's mouth and throat are clear. Tilt his head back slightly, pinch the nostrils shut, and force air from your lungs into his. When you remove your mouth from his, listen for air being exhaled.

Once normal breathing has started, you can stop the mouth-to-mouth, but keep a close watch.

If the victim has swallowed the insecticide, induce vomiting. Read the label first, however. Then give milk or water to help dilute what remains. Unless the doctor has advised you to do so, never give the victim any drugs or alcohol.

Most large cities have Poison Control Centers. These agencies handle all forms of poisoning. If you haven't already made a note of their phone number, look it up *now*. Keep it around, along with the numbers of your doctor, police and fire departments. If accidental poisoning happens, you certainly won't want to waste time trying to find the proper phone numbers. At the back of this book is a place for these numbers. Write them down, and keep the book handy in case of an emergency.

You'll probably never have to refer to it, or to the first aid information. With even a moderate degree of caution, every poison listed in this book is safe. Even if there is an accident with them, it is unlikely that the effects will be serious.

Still, be careful. And, *read the label.*

Questions on insects you may have can often be answered by a local university. The Agricultural Extension Division can be an invaluable aid to controlling local pests. Usually they are more than willing to cooperate.

Laws and regulations sometimes change. If you have any questions on recent regulations, use of pesticides, safety, and disposal of the empty container, refer to the nearest branch of the E.P.A. (Environmental Protection Agency).

Despite some of the seemingly silly rules handed down by the E.P.A., they do serve a useful purpose. It is their job to research the various pesticides and keep track of their properties.

Should you ever need any information from them, write to the region nearest to you.

E.P.A.Region I
Room 2303
John F. Kennedy Federal Bldg.
Boston, MA 02203

Region II
Room 1005
26 Federal Plaza
New York, NY 10007

Region III
6th Ave. & Walnut Street
Philadelphia, PA 19106

Region IV
1421 Peachtree St. N.E.
Atlanta, GA 30309

Region V
230 S. Dearborn
Chicago, IL 60604

Region VI
Suite 1100
1600 Patterson St.
Dallas, TX 75201

Region VII
Room 249
1735 Baltimore Ave.
Kansas City, MO 64108

Region VIII
Lincoln Tower Bldg., Suite 900
Denver, CO 90203

Region IX
100 California St.
San Francisco, CA 94111

Region X
1200 Sixth Ave.
Seattle, WA 98101

If you have an interest in doing further reading on insects, check your local library or bookstore. There are several excellent books available. My favorites are:

Pictorial Encyclopedia of Insects,
V. Stanek. Paul Hamlyn (1969).

Insects on Parade,
Clarence J. Hylander. Macmillan Co. (1957).

Handbook of Pest Control,
Arnold Mallis. MacNair-Dorland Co. (1954).

Emergency Phone Numbers

Local Poison Control _____

Paramedics (911) or _____

Doctor _____

Notes and Other Information

Index

Index